THE INCREDIBLE HISTORY OF
INDIA'S GEOGRAPHY

Currently the global strategist of one of the world's largest banks, Sanjeev Sanyal divides his time between India and Singapore. A Rhodes Scholar and an Eisenhower Fellow, Sanjeev was named Young Global Leader for 2010 by the World Economic Forum. He has written extensively on economics, environmental conservation and urban issues. His first two books, *The Indian Renaissance: India's Rise After a Thousand Years of Decline* and *Land of the Seven Rivers: A Brief History of India's Geography*, were published by Penguin in 2008 and 2012 respectively. This book is an adaptation of the latter.

Sowmya Rajendran writes for children across age groups, from picture books for tiny readers to young adult fiction. She is also a columnist with the school edition of *The New Indian Express* and Sify movies. Sowmya lives in Pune without any dogs or cats in the house.

From

Swapnil

THE INCREDIBLE HISTORY OF INDIA'S GEOGRAPHY

SANJEEV SANYAL

WITH

SOWMYA RAJENDRAN

PUFFIN BOOKS

An imprint of Penguin Random House

PUFFIN BOOKS

USA | Canada | UK | Ireland | Australia
New Zealand | India | South Africa | China

Puffin Books is part of the Penguin Random House group of companies whose addresses can be found at global.penguinrandomhouse.com

Published by Penguin Random House India Pvt. Ltd
7th Floor, Infinity Tower C, DLF Cyber City,
Gurgaon 122 002, Haryana, India

First published in Puffin by Penguin Books India 2015

18 17

ISBN 9780143333661

Typeset in Electra by Manipal Digital Systems
Printed at Replika Press Pvt. Ltd, India

www.penguin.co.in

Contents

1
Of Shakes and Quakes

If someone asked you to point out where India is on the world map, you'd probably do it in a jiffy. There it is, jutting into the Indian Ocean with Sri Lanka forming a teardrop beneath its land mass. The image is a very familiar one. But what if you were told that the Indian subcontinent was not always located where it is today? That it was once attached to Africa and Madagascar?

This is a fairly new discovery. For a long time, till the early twentieth century, people thought that continents were fixed land masses. But in 1912, a geologist called Alfred Wegener came up with the theory of continental drift.

> Continental drift is the movement of the continents across the ocean bed. Now don't look down at your feet to see if you are moving—this drifting happens very, very slowly, over hundreds of millions of years!

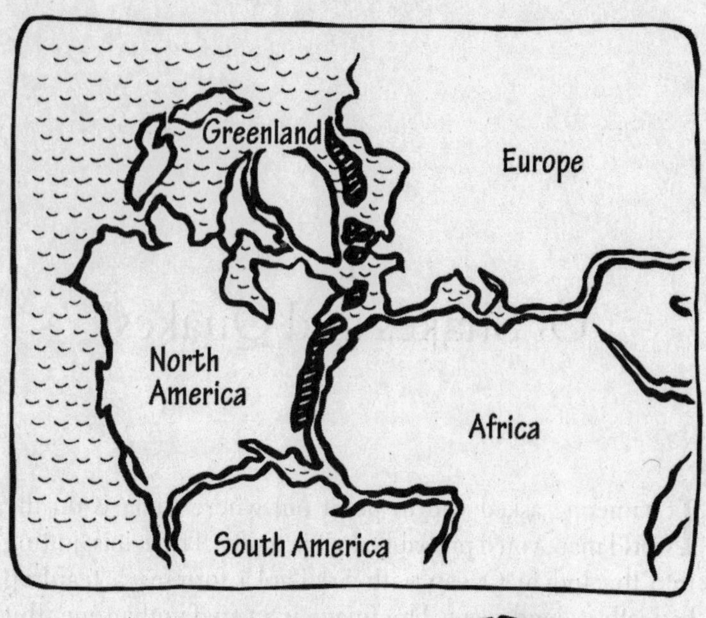
Greenland
Europe
North America
Africa
South America

TOLD YOU.
Alfred Wegener

Wegener expanded on this idea in his book *The Origin of Continents and Oceans*, which was published in 1915. He argued that the present continents all came from one single land mass that later drifted apart. While this sounded strange to people at that time, it explained why the world map looks like a jigsaw puzzle with different countries and continents appearing like they could fit into each other. These countries are far apart but their outlines seem like they could be joined together.

It took nearly fifty years for Wegener's arguments to be scientifically proved! In the late fifties and sixties, a great deal of new geological data established what Wegener had suspected: the earth's crust is a patchwork of plates and these plates are moving relative to each other. This led to the modern theory of plate tectonics.

Here is how scientists believe it all happened . . .

A billion years ago, there was a supercontinent called Rodinia. It was probably located south of the equator but we are still not sure about its exact shape or size. This supercontinent broke up around 750 million years ago and the various pieces, i.e. continents began to drift apart. This period is loosely called the Pre-Cambrian period. There were only single-cell organisms like bacteria alive then.

Did you know?

The Aravalli Range in India is thought to be the oldest surviving geological feature anywhere in the world! These mountains were once very tall, maybe as tall as the Himalayas, but over hundreds of millions of years, they have been eroded down to low hills and ridges. The northernmost point of the Aravallis is the North Ridge near Delhi University. Farther south, near the Gujarat-Rajasthan border, these short hills turn into mountains again. The Guru Shikhar peak at Mount

Abu rises to 1722 metres above sea level and is considered to be a sacred place. The Rajput warrior clans claim that their ancestors arose from a great sacrificial fire on this mountain! Despite the significance of the Aravallis, they are under threat today because of reckless mining and quarrying.

Fossil records show that around 530 million years ago, there was a sudden appearance of a large number of complex organisms on the earth. This is called the Cambrian Explosion—but remember that we're talking in geological terms. This 'explosion' took millions of years to happen. Over the next 70–80 million years, a whole new array of life forms evolved. While all of this was happening, the continental land masses began to reassemble and, about 270 million years ago, fused into a new supercontinent called Pangea.

How did the new world look? As you can see, the Indian craton is wedged between Africa, Madagascar, Antarctica and Australia.

A craton is a large, stable block of earth which forms the centre of a continent.

It was on Pangea that the dinosaurs appeared 230 million years ago. But the earth was still restless and Pangea began to break up around 175 million years ago, during the Jurassic era. It first split into a northern continent called Laurasia (consisting of North America, Europe and Asia) and a southern continent called Gondwana (Africa, South America, Antarctica, Australia and India). You might have heard of the Gond tribe of central India—well, this is where the name comes from!

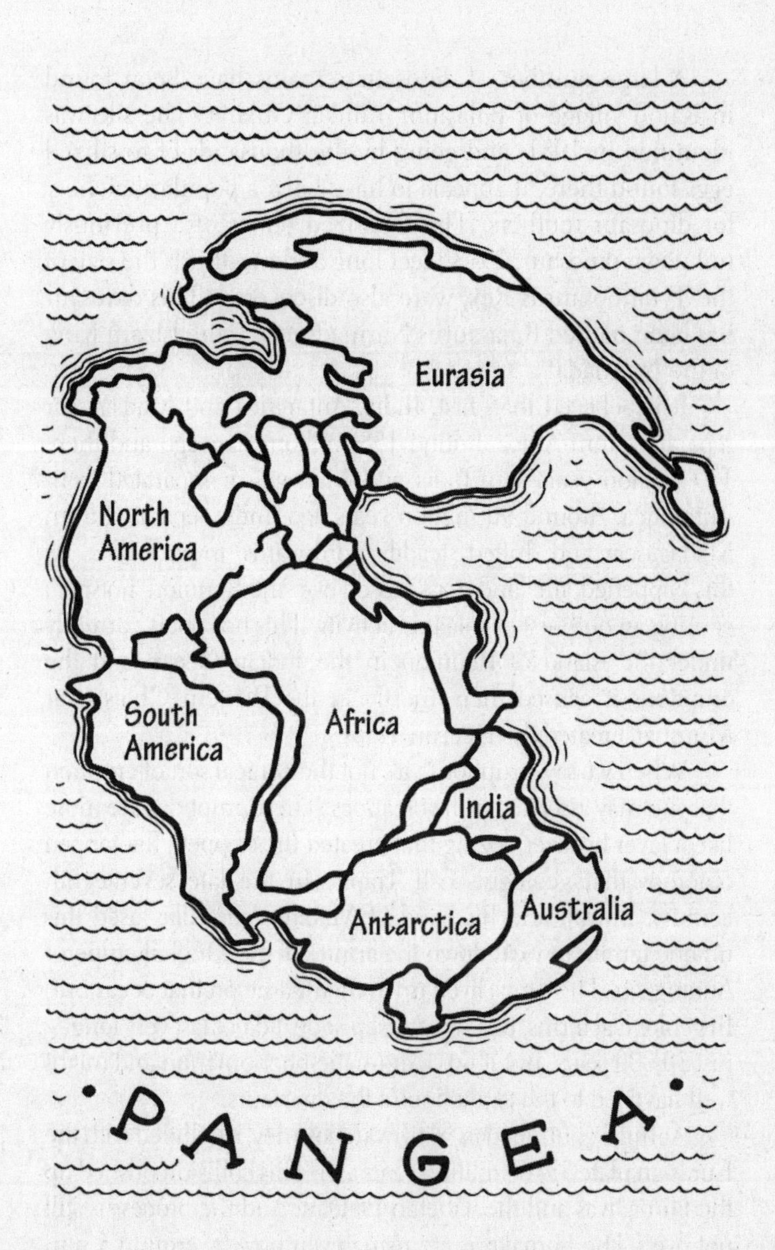
Eurasia
North America
South America
Africa
India
Antarctica
Australia
PANGEA

A large number of dinosaur remains have been found in Raioli village of Balasinor Taluka, Gujarat. The site was identified in 1981, and going by the thousands of fossilized eggs found there, it appears to have been a popular hatchery for dinosaur mothers. The fossilized bones of a previously unknown dinosaur, 25–30 feet long and two-thirds the size of the Tyrannosaurus Rex, were also discovered. This dinosaur has been named Rajasaurus Narmadsensis—the Lizard King of the Narmada!

It is believed that, first, India, Antarctica and Madagascar separated from Africa around 158 million years ago and then, 130 million years ago, India and Madagascar separated from Antarctica. Around 90 million years ago, India separated from Madagascar and drifted steadily northwards, towards Asia. As this happened, the land mass passed over the Reunion 'hotspot', causing an outburst of volcanic activity. This hotspot is currently under the island of Reunion in the Indian Ocean and the eruptions it caused then, mostly in the Western Ghats near Mumbai, created the Deccan Traps.

When we say 'eruptions', it's not the conical sort of eruption that you may associate with volcanoes. These eruptions are more like a layer-by-layer oozing that created the stepped, flat-topped outcrops that geologists call Traps. (In the late seventeenth century, Shivaji and his band of Maratha guerrillas used this unique terrain to wear down the armies of the Mughal emperor Aurangzeb. The Traps lived up to their name on that occasion!) In geological terms, this volcanic episode did not last very long—just 30,000 years. But it was a dramatic phenomenon and might well have led to the extinction of the dinosaurs.

As India continued its northward journey, it collided with the Eurasian plate 55–60 million years ago. This collision pushed up the Himalayas and the Tibetan Plateau. And the process is still not over! The Himalayas are rising even now by around 5 mm every year, although erosion reduces the actual increase in height. This region is considered to be seismically unstable, meaning that it is prone to frequent and powerful earthquakes.

Did you know?

What are now the towering Himalayan mountains were once under the sea. This is why marine fossils are commonly found high up in the range.

While most of the above is generally accepted by geologists, there are many unresolved issues and findings that don't tie in with this story. For example, a large number of insects preserved in amber were discovered in Vatsan, 30 km north of Surat, in a geological zone called Cambay Shale. About 700 species of insects, representing fifty-five families, were found. But these insects were not unique to India. They were similar to those found in other countries in other continents, as far away as Spain. If we are to go by the currently accepted view about the northward drift of the Indian land mass, we have to believe that the subcontinent was an isolated island for tens of millions of years. But if these insects emerged then, how did they come to India? Were there other islands that allowed them to hop across to the subcontinent? Maybe the Indo-Asian collision happened earlier than what we think? We really don't know!

Nonetheless, India continued to push into Asia, making the subcontinent tectonically very active. This meant that there were many powerful earthquakes that took place during this time. This region is still very unstable. In 2005, an earthquake in North Pakistan and Pakistan-occupied-Kashmir registered a magnitude of 7.6 on the Richter scale and claimed 80,000 lives (note that the Richter scale is a logarithmic scale, so each point increase is equivalent of a ten-times increase in the amount of shaking and 31.6 times the amount of energy released).

There have been many far more powerful earthquakes that have been recorded along the mountain range. The Assam earthquake of 1950 registered a magnitude of 8.6 and is one of the most powerful earthquakes ever recorded. It happened in a sparsely populated area and yet killed 1500 people. Imagine if it had taken place in a densely populated area—the lives of millions of people would have been in danger. This is why the Himalayan range is one of the most dangerous places to build large dams.

If the Aravallis are one of the oldest geological features, the Gangetic plains are among the youngest. They started out as a marshy depression running between the Himalayas and an older mountain range called the Vindhyas. Silt brought down by the Ganga and its tributaries began to fill up this hollow and create a fertile alluvial plain. The Ganga changed course repeatedly and shifted southward leaving behind **oxbow** or curved lakes that can still be seen. Early humans would have seen it all happening. The Ganga continued to drift southward and was arrested only when it nudged into the Vindhyas near Chunar (close to Varanasi). It is the only place in the plains where a hill commands such a view over the river. And that is why the Chunar fort was considered a strategic location in the times of warring kingdoms. It was once said that he who controlled the Chunar fort also controlled the destiny of India!

A walk through the fort is a walk through Indian history. The walls resonate with the tales of the legendary King Vikramaditya, the Mughals, Sher Shah Suri and Governor-General Warren Hastings. There are remains here from each era, including an eighteenth-century sundial. There are British graves below the walls, too. You must be familiar with the national emblem of India, of course.

ASHOKAN PILLAR

These are the Mauryan lions of Sarnath. They were carved out of the stone quarried from the south-west of the Chunar fort. We will return to them in Chapter 3.

Move It, People!

Many people assume that the similarities between present-day Indian and African mammals are because India was once attached to Africa. Elephants, rhinos and lions are common to both. But, as we have seen, India separated from Africa during the dinosaur era. So actually, these big mammals came to India because of its geographical reattachment to Eurasia and the changing climate zones that allowed or forced these animals to migrate.

A genetic study of the frozen remains of a Siberian mammoth that died 33,000 years ago revealed that the Asian elephant is more closely related to the mammoth than to the African elephant! It appears that the genetic lines of the Asian and the African elephants separated six million years ago whereas the Asian elephants and the mammoths diverged only 4,40,000 years ago.

Many Indian animals also came to the subcontinent from the east. The tiger is one such example. Some say that the tiger came from Siberia while others say it came from South China. Two-million-year-old remains of the tiger's ancestors have been found in Siberia, China, Sumatra and Java but it's a relative newcomer to India. The Bengal tiger is believed to have come to India only about 12,000 years ago.

Where were the human beings when all of this was happening? Most scientists agree that human beings first evolved in Africa around 2,00,000 years ago. The San tribe of the Kalahari (also called the Bushmen) is probably the oldest surviving population of humans. A genetic study of the members of this tribe revealed that they show the greatest genetic variation of any racial group. This means that they are likely to be the direct descendants of the earliest modern human population.

What do we mean by 'modern humans'? Human beings, as we call ourselves today, are only one kind of hominids (the genetic classification of which humans are a part) to have walked the earth. More than a million years ago, pre-modern humans like Homo erectus used stone tools and had wandered as far as China and Java. When modern humans were evolving in Africa, their close cousins, the Neanderthals, were already well established in Europe and West Asia.

We are survivors from a large family tree. There were many challenges that modern humans had to meet in those times. The first attempt by modern humans to leave Africa was a failure. Archaeological remains in the Skhul and Qafzeh caves in Israel show that modern humans may have made their way to the Levant (the region immediately east of the Mediterranean) about 1,20,000 years ago. The planet was then enjoying a relatively wet and warm interglacial period, which would have allowed them to wander up north. However, this climatic period didn't last for long and a new ice age started. It looks like the early settlers who made it to this point either died out or were forced to go back. The Neanderthals who were better adapted to the cold probably reoccupied the area.

An ice age is a long period of time when the temperatures on the earth are so low that the ice covering the surface—glaciers, polar ice caps, continental ice sheets—expand. In the history of the earth there have been many ice ages that alternate with warm periods, when the ice melts, the sea levels rise and the climate is warmer. Just to make it a little confusing, though, within the time period of an ice age you also have shorter periods of warmer and colder temperatures that alternate! The colder periods are called glacials, because the glaciers grow, and the warmer periods are called interglacials.

For the next 50,000 years, our ancestors remained in Africa. Around 65,000–70,000 years ago, a very small number, perhaps a single band, crossed over from Africa into the southern Arabian peninsula. And it was from this group that all non-Africans descended!

Climate and environment had a very big impact on the expansion of modern humans. Our planet goes through natural cycles of cooling and heating. When the modern humans made their way out of Africa, the earth was much cooler and much of the world's water was locked in giant ice sheets because of the low temperature. As a result, the sea levels were as much as 100 metres lower than today and coastlines and climate zones were very different, too. The early band of humans migrating from Africa to southern Arabia would have had to make a relatively short crossing across the Red Sea. They would have also found the Arabian coastline to be wetter and better for survival.

After this, the modern humans made their way along the coast to what is now the Persian Gulf. The average depth of the Persian Gulf is just 36 metres. With sea levels 100 metres below current levels, this area would have been a lush and fertile plain. It would have been paradise for the modern humans who are likely to have flourished and increased their numbers. Central Asia and Europe would have been very cold at this time because of the ice age. The modern humans must have spread out along the Makran coast into the Indian subcontinent.

At some stage, groups of the Persian Gulf people explored the Indian subcontinent more. But they weren't the first to do this. The Neanderthals from Europe steadily moved westwards till one of their last bands died out in a cave in Gibraltar. But we don't really know what happened to the pre-modern hominids of Asia.

Was it the eruption of the Toba volcano in Sumatra 74,000 years ago that led to the extinction of the pre-modern hominids of Asia? Excavations have shown that peninsular India was covered in volcanic ash from the eruptions. Experts still disagree on what really happened because of these eruptions but it's possible that they led to the disappearance of the pre-modern hominids, clearing the way for the modern humans.

The modern humans who had reached the subcontinent spread quickly through it and then to South East Asia. Some believe that the indigenous tribes of the Andaman and Nicobar Islands were maybe descendants of the earliest people who came into the region!

From here, one branch reached Australia around 40,000 years ago and became the ancestors of the aborigines. Studies have confirmed that the Australian aborigines have a genetic link with aboriginal tribes in South East Asia. However, for a long time, researchers couldn't find a direct genetic link between present-day Indians and native Australians. But in 2009, a study published by the Anthropological Survey of India found genetic traces to link some Indian tribes with native Australians. These were very tiny traces but still, they were there! The researchers suggested that the Indian and Australian groups had separated about 50,000–60,000 years ago.

We've talked about the adventurous people who left the Persian Gulf and went exploring. But what of those who were content to stay behind? The population that remained in the vicinity of the Persian Gulf and the subcontinent stayed there for several thousand years. Scientists think that many important genetic branches came from this area at this time. During the relatively warmer interglacial periods, sub-branches would have spread farther out into Europe, Central Asia and so on. But you have to remember that temperatures would have still been far lower than present-day levels and that there would have been many drastic climatic changes. Much of the Persian Gulf is now underwater, so it's not very easy to conduct research on the people who lived there.

This is a very short and simplified account of what happened over tens of thousands of years. We're talking about very small Stone Age bands of fifty to hundred people over vast expanses of time and space. Their movements would not have always been systematic. They might have wandered somewhere, come back, gone to places that didn't lead anywhere and so on. Just as there were groups coming into the subcontinent, there were others that were going out. Scientists think that India may have been the source of a number of genetic lineages that can now be traced worldwide.

Natural calamities, hunger, tribal wars and disease would have decided which of these groups survived and which of them didn't. There are plenty of remains of these early humans in Stone Age sites scattered across India. Bhimbetka in central India is one of the most extensive sites in the world. The hilly terrain is littered with hundreds of caves and rock shelters that appear to have been inhabited almost continuously for 30,000 years! It is now a UNESCO World Heritage Site.

Did you know?
There were once ostriches in the Indian subcontinent! Archaeologists have found beads and ornaments made from ostrich eggshells in Stone Age sites. Was it the Stone Age fashion industry that led to the disappearance of the bird?

The last full-blown ice age started around 24,000 years ago, reached its peak around 18,000–20,000 years ago and then warmed up. Around14,000 years ago, the ice sheets began melting rapidly, the sea levels were rising around the world and weather patterns were changing. The Persian Gulf began to fill up 12,500 years ago. Around 7500–8000 years ago, the Gulf Oasis was completely flooded. Is this the event that is referred to as the Great Flood in Sumerian and Biblical accounts? It's quite possible!

Recent archaeology suggests that the people of the Persian Gulf moved to higher ground around 7500 years ago. They also seem to have learned how to travel by water. A small clay replica of a reed boat and a depiction of a sea-going boat with masts from this period have been found in Kuwait. By this time, people knew how to farm, domesticate animals and build boats. Some groups made their way into Central Asia, taking advantage of the warmer temperatures. Others might have made their way into Europe where earlier migrations had previously pushed out the Neanderthals. Groups from South East Asia had already established themselves in China and the warmer climate would have allowed them to expand.

The Indian coastline moved several kilometres inland to roughly resemble what we would now recognize on the map. The sea moved inland all along the coast and there were two places where very large land masses were flooded. One was where we now have the Gulf of Khambat (Cambay), just south of the

Saurashtra peninsula of Gujarat. The other land masses extended south from the Tamil coast and would have included Sri Lanka.

> In Indian mythology, one of the ten avatars of Vishnu, the Protector, is that of the fish. It is said that Vishnu took the form of a fish (Matsya) and warned Manu, the legendary king, about a great flood that would threaten all life. Manu built a large ship and filled it with seeds and animals. Matsya then towed the ship to safety. Doesn't this remind of you of Noah's Ark? Are these legends a memory of the ancient floods?

In 2001, marine archaeologists found two underwater locations in the Gulf of Khambat. They seem to be the remains of large settlements that would have been flooded about 7500 years ago. Scholars are still finding out the exact nature of these discoveries, but if proved, they would be truly remarkable. Though we don't know about these for sure yet, it is reasonable to say that the changes in weather patterns and the sharp rise in sea levels must have made people in those times move from one settlement to another.

Earlier, it was thought that people from the Persian Gulf area carried the knowledge of farming to other regions. There is evidence to show that some of the crops that were farmed systematically in the subcontinent, around 7000 years ago in Mehrgarh, Baluchistan, were West Asian species such as wheat and barley. Did this mean that Indians learned to farm from West Asian migrants and only later managed to domesticate local plants such as eggplant, sugar cane and sesame? But recently, researchers have uncovered evidence that Indians may have independently developed farming, including the cultivation of rice. Did the knowledge of farming travel from one region to another or did different groups develop it independently in around the same time? The evidence now suggests parallel development.

What we do know is that by the end of the Neolithic age, there was a fairly large population living in India. Who were these people? How are present-day Indians related to them?

What Do Ya Mean, Gene?

Up to the early twentieth century, it was believed that India was inhabited by aboriginal Stone-Age tribes till around 1500 BCE when Indo-Europeans called 'Aryans' invaded the subcontinent, bringing with them horses and iron weapons. Indian civilization was seen as a direct result of this invasion. Though this theory didn't have any solid evidence to back it, it became a popular explanation for why Indian and European languages have similarities. It was also politically convenient at that time because it made the British colonizers appear as if they were merely latter-day 'Aryans' who'd come to further 'civilize' the local population.

The theory, however, took a beating when remains of the sophisticated Harappan civilization were discovered. These discoveries proved that Indian civilization was well underway even before 1500 BCE. But strangely, the 'Aryan invasion' theory was not thrown away. It was instead modified to suggest that a people called the Dravidians (supposed ancestors of modern-day Tamils) created these cities and that they were later destroyed by the invading Aryans. But this theory was also flawed because there is no archaeological or literary evidence of such a large-scale invasion. The Harappan cities did not suddenly collapse but suffered a slow decline because of environmental reasons.

India is a country with a bewildering mix of castes, tribes and language groups. Some of these groups came to India in historical times—Jews, Parsis, Ahoms, Turks to name a few. But there are also many populations that have lived in the country for a very long time. Many groups migrated to different parts of the country and settled there over thousands of years. So where a group is found today may not be where it came from originally. Over the years, most groups have mingled and yet a few have retained their unique identity even now—some of the tribes in the Andaman and Nicobar Islands and the North-Eastern states, for example.

What we have to remember when we study such a complex mix of people is that there are no 'pure' races. Indians come in all shapes, sizes and shades and these variations can be quite dramatic even within the same family! But there are some patterns of genetic distribution that we can see.

What is a gene?

A gene is the basic physical and functional unit of heredity. Do you have eyes like your mother? Is your nose like your father's? All of this came to you through your genes! Genes are sections of long chains of molecules called DNA (deoxyribonucleic acids) that give instructions to make molecules called proteins, which then build our bodies. Every person inherits two copies of genes, one inherited from each parent. Over time these genes mutate or change slightly. The accumulation of these mutations over long periods is responsible for evolution.

In 2006, there was a study that said India's population mix has been broadly stable for a very long time and that there has been no major injection of Central Asian genes for over 10,000 years. This means that even if there had been a large-scale influx of 'Aryans', it would have taken place more than 10,000 years ago, long before iron weapons and the domestication of the horse. The study also suggested that the population of Dravidians had lived for a long time in southern India and that the so-called Dravidian genetic pool may have even originated there.

Another study published in 2009 suggested that the Indian population can be explained by the mixture of two ancestral groups—the Ancestral South Indian (ASI) and the Ancestral North Indian (ANI). The ASIs are the older group and are not related to Europeans, East Asians or any group outside the subcontinent. The ANIs are a somewhat more recent group and are related to Europeans. The ANI genes have a large share in North India and account for over 70 per cent of the genes of Kashmiri Pandits and Sindhis. But the ANI genes also have a large 40–50 per cent share in South India and among some of the tribal groups of central India.

Is the ANI-ASI split same as the Aryan-Dravidian theory? Firstly, the ANI and ASI are not 'pure' races. They are just different genetic mixes, each of which contains many strands. The terms 'Aryan' and 'Dravidian', on the other hand, are not just about genetics; they also carry strong cultural connotations. For instance, the 'Aryans' are usually linked to the Vedic tradition while the 'Dravidians' are linked to the Sangam literary tradition. But we can't conclude that this is the same as the ANI-ASI framework because these two groups emerged well before the Vedic tradition, Sangam literature, or the Harappan civilization. We are talking about small bands of hunter-gatherers and early farming communities rather than the thundering war chariots, iron weapons and fortified cities that are said to have been part of an 'Aryan-Dravidian' rivalry.

Did you know?

Manu, the Indian Noah, was said to have been the king of the Dravidians before the flood but is repeatedly mentioned in the Vedic tradition as an ancestor!

As we shall see, climate change and the drying of a river caused these two groups to mix very rapidly from around 4200 years ago. Simply said, after thousands of years of mixing, Indians are very closely related to each other and it is pointless to try and find out who is more Aryan and who is more Dravidian. There are also many groups in India that don't fit in within the ANI-ASI framework and which have influences from other parts of the world. Genetics has just confirmed what we can see for ourselves—Indians are a mongrel lot who come in all shapes, sizes and complexions!

What about the genetic links of North Indians to Europeans? And how do we explain the linguistic similarities between Indian and European languages if we don't accept the 'Aryan-Dravidian' theory? When we talk about a genetic link between North Indians and some Europeans and Iranians, what we're usually referring to is a gene mutation called R1a1, and more specifically, a subgroup called R1a1a. This gene is common in North India and among East Europeans such as the Czechs, Poles and Lithuanians. There are smaller concentrations in South Siberia, Tajikistan, north-eastern Iran and in Kurdistan (that is, the mountainous areas of northern Iraq and adjoining areas). Interestingly, the gene is rare among Western Europeans, western Iranians and through many parts of Central Asia. But how is it that this gene is present in the Indian subcontinent and Eastern Europe while skipping Central Asia and Western Europe?

In 2010, it was discovered that the oldest strain of the R1a1a branch was concentrated in the Gujarat-Sindh-Western Rajasthan area, suggesting that this was close to the origin of this genetic group. European carriers of R1a1a also displayed a further mutation, M458, which is not found at all in their Asian cousins. Since the M458 mutation is estimated to be at least 8000 years old, the two populations must have separated before or during the Great Flood. Thus, the genetic linkages between North Indians and East Europeans are best explained by the sharing of a common ancestor, perhaps from just after the end of the last ice age. Does this also have to do with climate change? Maybe!

The most common gene in Western Europe is R1b. This is related to R1a1 and possibly also originated in the Persian Gulf area but the two separated a long time ago—probably during or before the last ice age. India has a relatively low concentration of R1b. Could we be dealing with two major genetic dispersals occurring from the Persian Gulf-Makran-Gujarat region at different points in the climatic cycle? One occurring at the onset or during the last ice age with R1b carriers heading mostly west and another occurring around the time of the Flood involving R1a1 carriers?

There is also reason to believe that some Indian tribes moved westward to Iran and beyond during the Bronze Age. We'll read more about that in the next chapter. Cultural linkages could have also happened because of trade. The spread of Indian culture to South East Asia in ancient times and the popularity of the English language in the postcolonial period show that it is possible for cultural exchanges to happen even without war or large-scale migration.

Is There a Lithuanian in Your Family?

The caste system is not unique to India. Throughout history, we have seen different versions of the caste system in Japan, Iran and even in Classical Europe. What is remarkable about the Indian caste system is that it has survived over thousands of years despite changes in technology, political conditions and religion. Despite strong criticism and opposition within Hindu tradition itself, it has continued to exist.

It was once thought that the caste system originated because of the Aryan influx and the imposition of a rigid racial hierarchy. However, genetic studies have shown a largely South Asian origin for Indian caste communities. They suggest that Indian castes are profoundly influenced by 'founder events'. This means that castes are created by an 'event'—when a group separates out for some reason and later turns itself into an endogamous tribe. That is, marriages are restricted to the 'tribe'. Over time, this process leads to a varied social environment of groups and subgroups, sometimes combining and sometimes splitting off. Because of this, we don't have a single unified population but a complex networks of clans. Recent studies suggest that intermarriage between different groups was fluid 1900–4000 years ago—coinciding with the mixing of the ANI and ASI. However, about 1900 years ago, intermarriage became less common and the castes became more exclusive.

There is a difference between the genetic reality and the rigid and strictly hierarchical 'varna' system of castes described in the Manusmriti (Laws of Manu).

Varna is the term for the four broad categories into which traditional Hindu society is divided. The four varnas, in descending order in the hierarchy, are:
the Brahmins: priests, teachers and preachers
the Kshatriyas: kings, governors, warriors and soldiers
the Vaishyas: cattle herders, agriculturists, businessmen, artisans and merchants
the Shudras: labourers and service providers

The Manusmriti is often used by scholars as the framework to understand the caste system. It now appears that the description of this rigid system may have been a scholarly idea and it may have never really existed. Instead, what we have is a very flexible society where people from different castes adapted easily to changing times by altering their social roles. Till 1900 years ago, these groups also seem to have commonly intermarried but even after they became strictly endogamous, the status of different groups was fluid. For example, if a new group has to be accommodated, a new caste can be created. Similarly, a group can be promoted or demoted in status according to social conditions. This fits with what we know from historical experience, such as the emergence of the Rajputs in medieval times. In the past, it was advantageous for groups to move forward in the pecking order. But now, we have groups trying to be classified as 'backward' in order to benefit from affirmative action! The logic of both processes is the same.

Affirmative action is the policy of creating special provisions for people who belong to groups that have suffered from discrimination in some form. The reservation policy in India for certain caste groups is one such example.

2
Hello, Harappans!

Much of what we know about India's early history comes from two very different sources, but archaeologists and historians are not quite sure how they fit together. On one hand, there is the archaeological evidence of the sophisticated cities of the Harappan Civilization. On the other hand, there is the literature of the Vedic tradition. Both are roughly from the same geography and timeline and we will listen to both the tales separately.

Though the two sources are different, there is one thing that they both agree on: the drying of a great river that the Rig Veda calls the Saraswati. No matter which way we look at it, the drying of this river was an important geographical event that defined early India.

Blast from the Past

When the Lahore-Multan railway line was being built in the late nineteenth century, wagonloads of bricks for ballast were removed from some old mounds. The bricks were of very good quality and most people assumed that they must be from modern times. However, it was discovered that these bricks were, in fact, from a very old civilization, just like the Sumerians, the Minoans and the ancient Egyptians. This civilization was named the Indus Valley or Harappan Civilization.

Soon, more and more such sites were discovered. It took so long to discover the Harappan Civilization because they did not have grand structures like the Pyramids of Giza or huge palaces and temples that immediately arrest attention. The Harappans did have large buildings but we don't know what they were used for. However, the Harappan Civilization is truly remarkable because of its urban design and active municipal management. These discoveries challenged the old theory about 'Aryan' invasions introducing civilization to India.

One of the large buildings from Mohenjodaro, a site in Sindh, has been identified as the Great Bath. But we don't really know if the structure was used for religious rituals, as a bathing pool for the royal family, or for some other purpose altogether!

We see meticulous town planning in every detail—standardized bricks, street grids, covered sewerage systems and so on. Similarly, a great deal of effort was put into managing water. Mohenjodaro alone may have had 600–700 wells! One of the bigger cities, it must have had a population of around 40,000–50,000 people. Not all cities had the same solutions to the same problems. At Dholavira in Gujarat, water was diverted from two neighbouring streams into a series of dams and preserved in a complex system of reservoirs. Many houses, even the small ones, had their own bathrooms and toilets connected to a drainage network that emptied into soak jars and cesspits. The toilet commodes were made from big pots sunk into the floor.

Did you know?

These ancient toilets came equipped with a 'lota' for washing up. Though we no longer use the same toilet design in our homes, the lota has survived in Indian toilets!

Can You Read Harappan?

Dholavira is a good example of a large Harappan urban centre. It is on an island in the Rann of Kutch. At the centre of the settlement is a 'citadel', which consists of a rectangular 'castle' and a 'bailey' (the outer wall of the castle). The citadel must have contained the homes of the rich as well as public buildings. The castle, which is the oldest part of the city, was heavily fortified with thick walls and equipped to withstand military attack. Early scholars who studied the Harappan Civilization believed that they were uniquely peaceful and that there were no signs of military activity. Then why did they require such walls?

In front of the citadel, there is a large open ground that could have been used for many purposes—military display, sport, royal ceremonies or maybe the annual parading of the gods. Archaeologists have found tiered seating for spectators along the length of the ground.

Beyond the ceremonial grounds was the planned area where the common citizens lived. This division into a Citadel and Lower Town is quite common in larger Harappan settlements. As the city grew, more and more people began to migrate into it and these migrants could not be accommodated in the planned city. So what did they do? They settled down just to the east of the original Lower Town—forming a 'slum' area, so familiar to many of our big cities today! However, the political leadership of Dholavira responded to the situation. They expanded the urban limits and included the slums into the city. The slums were redeveloped and the Harappan municipal order was

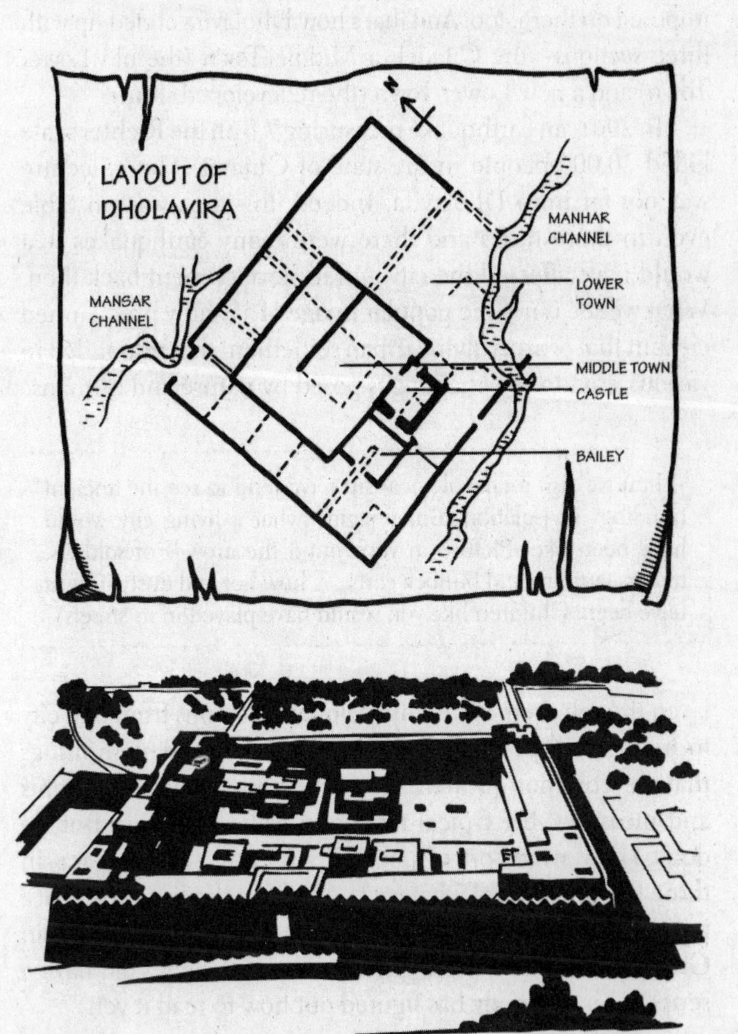
LAYOUT OF DHOLAVIRA
N
MANHAR CHANNEL
MANSAR CHANNEL
LOWER TOWN
MIDDLE TOWN
CASTLE
BAILEY

imposed on them, too. And that's how Dholavira ended up with three sections—the Citadel, a Middle Town (the old Lower Town) and a new Lower Town (the redeveloped slum).

In 2001, an earthquake measuring 7.8 on the Richter scale killed 20,000 people in the state of Gujarat. The epicentre was not far from Dholavira. Indeed, this area was unstable even in those times and there were many earthquakes that would have affected the city and its development back then. What we see is not the popular image of a rigidly pre-planned city but that of an evolving urban settlement that responded in various ways to the challenges posed by nature and humans.

> When we visit archaeological sites, we tend to see the ancient buildings in isolation. But imagine what a living city would have been like! Picture in your mind the crowds of soldiers, traders, artisans and bullock carts . . . how hot and dusty it must have been. Children like you would have played in its streets!

Even though there are many regional variations from one city to another in the Harappan Civilization, there are many things that are common to them. How they used standard weights and measures, the typical terracotta seals and so on. But we don't know what sort of political structure was in place in those times. Much of what we know about the historic events, political leaders, religion and language from the Harappan Civilization remain mere guesses. The Harappans did have a script . . . but nobody has figured out how to read it yet!

The Merchants of Meluha

We don't know much about the political history of the Harappans but we do know a lot about its geography. Over the

The Harappan Script

last century, thousands of sites have been found and several new sites are being discovered every year. It looks like a lot of people lived in the subcontinent even at this early stage.

The core of the Harappan Civilization extended over a large area, from Gujarat in the south, across Sindh and Rajasthan and extending into Punjab and Haryana. Many sites have been found outside the core area, including some as far east as Uttar Pradesh and as far west as Sutkagen-dor on the Makran coast of Baluchistan, not far from Iran. There is even a site in Central Asia called Shortughai along the Amu Darya, close to the Afghan-Tajik border.

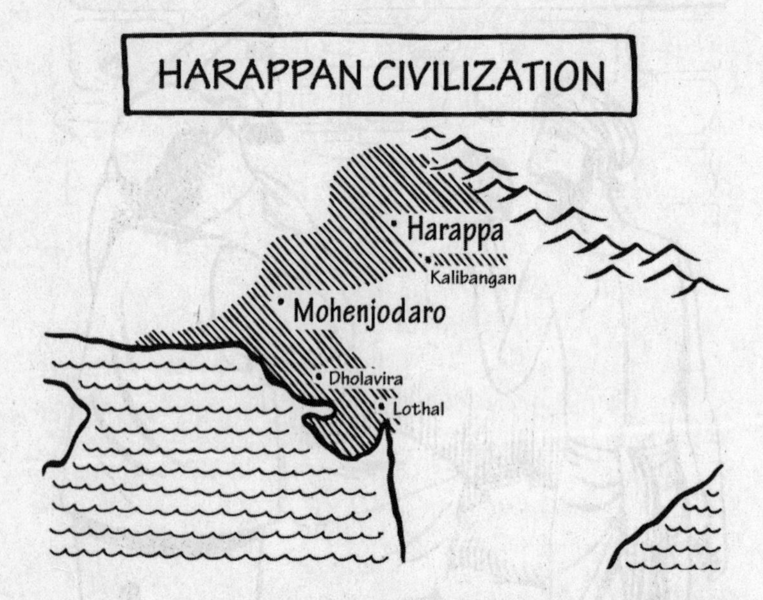

This extensive geographical spread means that the Harappan Civilization was made up of far more people than contemporary Egypt, China or Mesopotamia! What the Harappans lacked in grand buildings, they made up for in the sheer scale of their spread and the sophistication of their cities.

From what we know about the Harappans, they were actively engaged in domestic and international trade. For land transport, they used bullock carts. Cart ruts from Harappa show that even the axle-gauge of these carts was almost exactly the same as those used in Sindh today. The streets of the big cities would have been full of these carts ferrying merchants and their goods.

Did you know?

Traffic jams aren't exactly a recent phenomenon! The French traveller Tavernier spoke of how seventeenth-century Indian highways were clogged by bullock-cart caravans that could have as many as 10,000–12,000 oxen. When two such caravans met on a narrow road, there would be a traffic jam that could take two or three days to clear! The Harappan highways in those times would have been quite similar.

There were many rivers in this region and this meant that goods and people could be ferried from one place to another by waterways. A dry dock has been discovered at Lothal, near Ahmedabad, in Gujarat. The dock, which seems to be the world's first, is an impressive structure. It was connected by a canal to the estuary of the Sabarmati river and a lock-gate system was used to regulate water flow during tides. Next to the dock are the remains of the warehouses.

An estuary is a partly enclosed coastal body of salt water with one or more rivers or streams flowing into it. It also has a connection to the open sea.

There is strong evidence to show that the Harappans traded actively with the Persian Gulf. The merchant ships probably went along the Makran coast, perhaps with a pit stop at Sutkagen-dor and then sailed on to the ports of the Persian Gulf. Mesopotamian tablets mention a land called Meluha that exported bead jewellery, copper, wood, peacocks, monkeys and ivory—goods that sound like Indian exports. It's also likely that the Harappans exported cotton because they were the pioneers in the spinning and weaving of cotton. Even now, the Indian subcontinent is a major exporter of cotton textiles and garments.

But what did the Harappans bring to their land from other civilizations? We don't know! Hardly any object of Mesopotamian origin has been found at the Harappan sites. Did they import consumable goods like dates and wines? We don't know what they bought from Iran and Central Asia either. Archaeologists have found a Harappan outpost in Shortughai on the Afghan-Tajik border. What were the Harappans doing there? Could they have gone there to buy horses? Indians have always had problems with breeding good-quality horses—even Marco Polo commented about this in the thirteenth century! We know that as late as the nineteenth century, Indian rulers imported large numbers of horses from Central Asia and Arabia . . . but we'll talk more about this later.

What Happened to India's First Cities?

We now know that this civilization did not suddenly appear or disappear. Rather, these cities were built gradually, sometimes rebuilt on older sites, and their disintegration too was gradual. But why were these cities abandoned? This did not happen overnight, so it's clear that it wasn't because of 'Aryan' invasions as it had been thought earlier.

The evidence points to the wrath of nature. A number of studies have shown that the area which is today the Thar Desert was once far wetter, and that the climate slowly became drier. It is possible that the process of drying had already begun during the Mature Harappan period (2600 BCE to 2000 BCE). Around 2200 BCE, the monsoons had become weaker and there were prolonged droughts. This was a widespread phenomenon that also affected Egypt and Turkey. Poor monsoons and droughts would have created an agricultural crisis for a heavily populated region but the Harappans were faced with an even bigger problem—the drying up of the river system on which the civilization was based.

Most of the settlements of the civilization were around a river that we now know as the Ghaggar—not the Indus as widely believed. The Ghaggar is now little more than a dry riverbed that contains water only after heavy rains. However, surveys and satellite photographs confirm that it was once a great river that rose in the Himalayas, entered the plains in Haryana, flowed through the Thar-Cholistan Desert of Rajasthan and eastern Sindh and then reached the sea in the Rann of Kutch in Gujarat.

The Rann of Kutch has a very strange marshy landscape. This is partly due to the fact that it was once the estuary of a great river. Much of it is now dry desert but satellite photographs show that there is still a substantial amount of underground water along the old channels. Wells, even drilled at shallow depths, give fresh water in the middle of the Thar Desert!

The Ghaggar emerges from hills just east of Chandigarh and is joined by a number of other seasonal rivers in the plains of northern Haryana. The Ghaggar and some of these rivers were perennial in ancient times. That is, they always had water, no matter what the season. Satellite images show that both the Sutlej and the Yamuna once flowed into the Ghaggar—this means that it would have been a truly mighty river!

However, at some point, the Ghaggar seems to have lost its main sources of glacial melt from the Himalayas. The Sutlej and the Yamuna, its largest tributaries, abandoned it for the Indus and the Ganga respectively. Once again, this seems to have happened because of tectonic shifts. The Ghaggar no longer flowed to the sea. It may have struggled on with the help of seasonal tributaries but even these failed as the climate changed.

All of this would have taken place over decades or even centuries and different parts of the Harappan world would have experienced these changes differently. Cities on the banks of the Indus, for example, may have suffered floods as waters from the Sutlej suddenly entered their region. The Pakistan floods of 2010 provide a glimpse of what this might have felt like—especially if such an event had caused the mighty Indus to shift course.

What impact did the drying of the Ghaggar have on the Harappans? The climate was wetter when the Ghaggar was in full flow in the early phase of the civilization. There is evidence to suggest that urban centres actually flourished when the Ghaggar began to dry up—there is a dense concentration of Harappan sites in the Thar Desert around the time we think that the Ghaggar might have started to dry up. Maybe the drying weather briefly created conditions that allowed them to flourish.

However, around 2000 BCE, conditions worsened. The lack of water began to affect the Harappans. Their carefully managed cities began to fall apart and they began to migrate. Too little water or too much water still causes people to sometimes migrate from their place of origin. Imagine the long lines of bullock carts, heavily laden with personal belongings, people leaving their old villages and cities in search of a better future!

In the north, the Harappans moved north-east to the Yamuna and Ganga. In Gujarat, the cities in Kutch were abandoned in favour of new settlements in the Narmada and Tapti valleys to the south. The later Harappan sites did have cultural connections with the old ones but they remained small settlements. The old urban sophistication had broken down.

Where Did the Harappans Go?

Even though some say that the Harappan culture disappeared with the disintegration of its cities, some others put forth compelling evidence to show that many of their cultural traits have been passed on over the years. For example, Indians usually greet each other with the 'namaste'. It is a common way to show respect. Do you know that several clay figurines from the Harappan sites have their palms held together in a namaste, too? Not just that, there are terracotta dolls of women with red vermilion on their foreheads—even today, many Hindu married women apply 'sindur' on their foreheads, don't they? Still, even though all of this is very intriguing, we cannot be absolutely sure that the Harappans used these gestures and symbols in the same ways as we do now.

The Harappans had a standardized system of ratios, weights and measures, many of which are echoed 2000 years later in the *Arthashastra*, a manual on governance and political economy written in the third century BCE. Some of these measures and ratios were used in India till the twentieth century! It was only since 1958 that we started using the metric system.

It has long been known that the game of chess originated in India. Chess pieces that look a lot like the modern equivalents have been found in Harappan sites. Isn't it amazing that a game we play in our modern world was also played more than 4000 years ago? The streets of Kalibangan, a large Harappan

Reconstruction of a marketplace in Harappa

site in Indian Punjab, are laid out with widths in a progression prescribed in the *Arthashastra*. Perhaps this indicates that the Harappans didn't just disappear but that they live on amongst us? This is why it is no coincidence that genetic data on ANI-ASI mixing fits exactly with the period when the Harappans were migrating. This mixing led to what we now know as the Indian civilization. However, as we said at the beginning of this chapter, there is one other parallel source that we must turn to which gives us clues about the origin of civilization in the Indian subcontinent—the Vedic tradition.

Digging Through the Rig Veda

The Vedas are the oldest scriptures of the Hindu tradition. There are four books or Vedas—Rig, Sama, Yajur and Atharva. They consist mostly of prayers, hymns, and instructions on how to conduct rituals and fire sacrifices. They were composed and compiled over several centuries by rishis or poet-philosophers.

The Rig Veda is the oldest of the four and is organized in ten sections. It is the oldest book in the world and remains in active use. It's considered by Hindus to be the most sacred of texts and one of its hymns, the Gayatri Mantra, is chanted by millions daily even today.

The Rig Veda is composed in a very old form of Sanskrit. But how old? We don't know for sure. The dates vary from 4000 BCE to 1000 BCE. Dating it is no easy task since it was probably compiled over decades or even centuries and remained a purely orally transmitted tradition till the third century CE. However, it is clear that the Rig Veda belongs to the Bronze Age as it does not mention iron. The earliest possible mention of iron comes in the Atharva Veda, which was compiled many centuries later and talks of a 'krishna ayas' or 'dark bronze'. Since we know that iron was in use in India by 1700 BCE, this would roughly date the Atharva Veda. Perhaps the Rig Veda was compiled a few centuries earlier, no later than 2000 BCE and possibly a lot earlier.

Since the nineteenth century, the Rig Veda has been used to find out more about early Indian history. While the book is about religion and philosophy and does not concern itself with social and political conditions, it does give us an idea about Bronze Age society, its social customs, its material and philosophical concerns, its gods and its tribal feuds. But it's difficult to make out historical events from the hymns.

The geography of the book, though, is very clear. To the east, the book talks of the Ganga river, and to the west, of the Kabul river. It also talks of the Himalayan mountains in the north and the seas to the south (i.e. the Arabian Sea). This is a very well defined geographical area and roughly coincides with the Harappan world.

What's most interesting is that the Rig Veda speaks repeatedly of a great river called the Saraswati. It is described as the greatest of rivers. No less than forty-five of the Rig Veda hymns shower praise on the Saraswati! No other river or geographical feature has got so much importance—the great Ganga is barely mentioned and the Indus, although referred to as a mighty river, is not given the same amount of respect. The Saraswati, on the other hand, was considered to be the mother of all rivers. It was even called the 'inspirer of hymns'—it's quite possible that the Rig Veda was composed on its banks.

However, there is no living river in modern India that fits this description. Some historians say that the Saraswati was simply a figment of imagination. Others believe it is the Helmand river in Afghanistan. But why go to other sources when the Rig Veda itself describes the geographical location of the river? In the *Nadistuti Sukta* (Hymn to the Rivers), the major rivers are listed from east to west, starting with the Ganga. The hymn clearly places the Saraswati between the Yamuna and the Sutlej.

There is only one river that could fit this description—the Ghaggar! It seems very likely that the Rig Vedic people and the Harappans were dealing with the same river. Unlike later texts, the Rig Veda does not mention a drying Saraswati. It mentions clearly that the Saraswati entered the sea in full flow. This would suggest that the text was composed before 2600 BCE! The Rig Veda talks of poets and compositions from an even earlier age but these works have not survived. Could it be that this culture coincided with the early Harappans? Not everyone may agree with these conclusions but these are definitely possibilities.

TRUE OR FALSE?

Why do some find it difficult to believe that the Rig Vedic people and the Harappans were the same? One of the oldest arguments is that the Rig Vedic people were nomads from Central Asia who could not have built the sophisticated cities of the Harappan civilization. They claim that this is why the Rig Veda reveals little knowledge of India's geography beyond the North West. But then, the Rig Veda neither mentions an invasion nor does it provide any information about Central Asia. All we can understand from the text is that these people were living in the area that corresponds roughly to modern Haryana, western Uttar Pradesh and Punjab (including Pakistani Punjab). They also knew of the Himalayas in the north, the seas in the south, the Ganga to the east and eastern Afghanistan to the west. It is possible that they may have known about South India and Central Asia but the text doesn't make any mention of this.

The Rig Vedic people knew about agriculture and cities—ones where they lived and ones where their enemies lived. They were not wild nomads as has been suggested. It is true that the Rig Veda does not talk about municipal order but why would a religious text talk about sewage systems anyway?

The other argument is that the Rig Vedic people were iron-wielding 'Aryans' who were at constant war with their enemies called 'Dasas'—either Harappans or aboriginal tribes. The term 'Arya' is commonly used in Sanskrit literature but never in the racial sense. 'Arya' means a cultured or noble person—which means all groups are likely to refer to themselves as Aryan and to their enemies as non-Aryan. The use of the word in a racial sense occurs in ancient Iran and modern Europe, but not in India. Similarly, we can't automatically call a non-Aryan enemy a 'Dasa' because the greatest of the 'Aryan' chieftains mentioned in the Rig Veda is a Dasa himself: Sudasa, son of Divodasa (more on him later).

What the Rig Veda describes is a sort of mishmash of tribal feuds between clans. These people belonged to the Bronze Age because the mention of iron appears many centuries later. Iron smelting was developed in central India, which was rich in iron ore. How could the Rig Vedic people be iron-wielding Aryans who conquered India when iron technology was probably discovered in India and that too long after the Rig Veda was composed?

Did you know?

The Rig Veda frequently mentions the bull and the horse. Harappan art features the bull quite a bit but the horse is missing. However, we do know that the Harappans were aware of the horse. They had a trading outpost in Central Asia where horses were widely used. There are Stone-Age rock paintings and horse bones from the pre-Harappan era which have been discovered in central India. So there must have been horses even earlier in this region. Also, while the Harappan seals don't have the horse, two terracotta figurines that depict a horse-like creature have been found. The set of chess pieces found in Lothal has a piece that looks like a horse's head. Some say horse bones have been discovered from Harappan sites, but sceptics say they are of asses and donkeys, not horses.

It is possible that the Harappans were a multi-ethnic society, just like India today. The Rig Vedic people may have been part of this bubbling mix.

Let's now look at the drying up of the Saraswati—the one event that archaeology and the texts categorically agree upon.

What Happened to the Saraswati?

We've already seen that the Ghaggar is most likely the Saraswati river that the Rig Veda speaks about at great length. Texts of later times repeatedly talk about how the Saraswati dried up. The *Panchavamsa Brahmana* tells us that the river disappeared into the desert. There are many legends and folktales about the river's downfall. But what was the cause behind it?

We know that the Sutlej and the Yamuna were once tributaries of the Saraswati and that the Yamuna seems to have

MAP OF RIVER SARASWATI

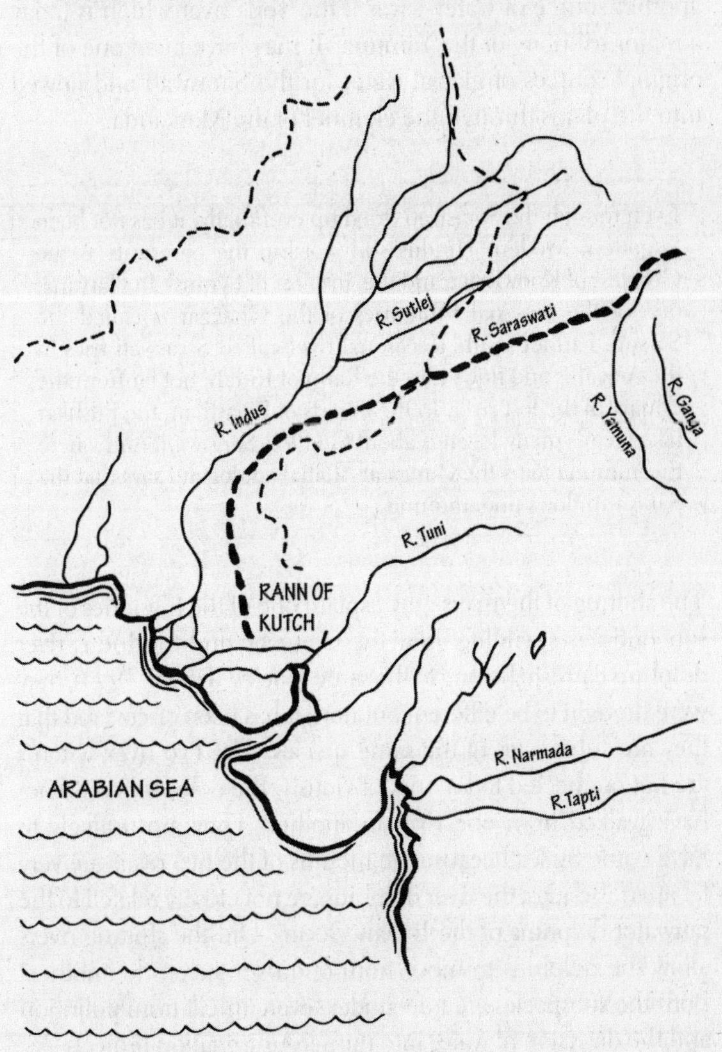

shifted its course because of a major tectonic event. The Sutlej, too, swung west towards the Indus. A Rig Vedic hymn hints at another source of water—was it the Tons river which is today a major tributary of the Yamuna? It may have been one of the original sources of glacial water for the Saraswati and flowed into the plains through the channel of the Markanda.

Even though the Saraswati dried up eventually, it has not been forgotten. Modern Hindus still worship the Saraswati as the Goddess of Knowledge and the 'inspirer of hymns'. In Haryana, one of the seasonal tributaries of the Ghaggar is called the Sarsuti. Farther south, a seasonal river called Saraswati rises in the Aravallis and flows into the Rann of Kutch, not far from the estuary of the lost river. In the deserts of Rajasthan, the Pushkar lake recalls many legends about Goddess Saraswati. And where the Yamuna joins the Ganga at Allahabad, legend says that the Saraswati flows underground.

The shifting of the rivers may explain one of the mysteries of the subcontinent's wildlife: how the Gangetic and the Indus river dolphins came to belong to the same species. Till the 1990s, they were thought to be different but now, it has been discovered that they are subspecies of the same species. The two river systems are not connected today and, obviously, the dolphins could not have walked from one river to another! They are unlikely to have come by sea because the mouths of the two rivers are very far apart. Besides, the river dolphins are not closely related to the saltwater dolphins of the Indian Ocean. Did the shifting rivers allow the dolphins to move from one river system to another? Both the subspecies are now under severe threat from pollution and the diversion of water into numerous irrigation projects.

Just as water problems plague our cities in modern times, people in that era must have also found it difficult to cope with the situation. The concern with water is echoed in the Vedas: Indra, king of the gods, is said to have defeated Vritra, a dragon who had held back the river waters behind stone dams. Indra slays Vritra after a great battle, destroys the dams, and sets the rivers free. What's more, the slaying of Vritra is specifically mentioned in a hymn praising Saraswati!

Land of the Seven Rivers

At the core of the Rig Vedic landscape was an area called Sapta-Sindhu (Land of the Seven Rivers). This was the heartland of the Rig Veda but the text does not clearly specify which seven rivers ran through it—as if it were too obvious and required no explanation. The hymns repeatedly describe the Saraswati as being 'of seven-sisters', so the sacred river must have been one of the seven but we're not sure which the others were. The conventional view is that the seven rivers include the Saraswati, the five rivers of Punjab and the Indus. This will mean that the Sapta-Sindhu region included Haryana, all of Punjab (including Pakistani Punjab) and even parts of adjoining provinces. A very large area!

But if we were to travel this terrain and read the Rig Veda several times, it's possible to reach another conclusion. The Vedas clearly mention a wider landscape watered by 'thrice-seven' rivers. We don't have to take it literally as referring to twenty-one rivers but it is obvious that the Sapta-Sindhu is only a part of the wider Vedic landscape. It's not likely that the Indus and its tributaries would have been part of the seven sisters as the Indus has long been considered a 'male' river in Indian tradition, after all! Could the Sapta-Sindhu refer only to the Saraswati and its own tributaries? Look at the following stanza:

Coming together, glorious, loudly roaring —
Saraswati, Mother of Floods, the seventh —

With copious milk, with fair streams strongly flowing,
Fully swelled by the volume of their waters.

It's possible to interpret this stanza to mean that the six rivers emptied into Saraswati, the seventh. There are several old river channels in the region, some of which still flow into the Ghaggar during the monsoon season. These include the Chautang (often identified as the Vedic river Drishadvati) and the Sarsuti. The Sutlej and the Yamuna were probably also counted among the Saraswati's sisters.

If we're right, it would mean that the Sapta-Sindhu was a much smaller area covering modern Haryana and a few of the adjoining districts of eastern Punjab and a bit of northern Rajasthan. This is the same area that ancient texts refer to as Brahmavarta—the Holy Land—where Manu is said to have re-established civilization after the flood. Is it a coincidence that the texts say that the Holy Land lay between the Saraswati and the Drishadvati, again roughly Haryana and a bit of north Rajasthan, but excluding most of Punjab?

But why was this small area given so much importance? The people of the Sapta-Sindhu must have been part of a culture that covered a much larger area. So what was so special about these seven rivers? Could it be because this region was the home of the Bharatas, a tribe that would give Indians the name by which they call themselves?

The Battle of the Ten Kings

Although the Rig Veda is concerned mostly with religion, there is one historical event that it mentions. This is often called the 'Battle of the Ten Kings', which occurred on the banks of the Ravi river in Punjab. It appears that ten powerful tribes ganged up against the Bharata tribe and its chieftain, Sudasa. This group appears to have mainly consisted of tribes from what is now western Punjab and the North West Frontier Province (both now in Pakistan). The Bharatas were an 'eastern' tribe, from what is now Haryana. Despite the combined strength of the ten powerful tribes, the Bharatas managed to crush them in battle. There are descriptions of how the defeated warriors fled the battlefield or were drowned in the Ravi.

The Rig Veda tells us that the Bharata warriors would have been dressed in white robes, each with his long hair tied in a knot on his head. There would have been horses neighing, bronze weapons shining in the sun and perhaps the rhythmic sound of Sage Vashishtha's disciples chanting hymns to the gods. The Saraswati would have been a mighty river then and it is likely that there would have been rafts ferrying men and supplies across the river. Just close your eyes and imagine the scene!

How did the Bharatas single-handedly defeat these tribes? The intelligence and military tactics of Sudasa and his guru Vashishtha must have played a role but it's also possible that it had something to do with access to superior weapons. The territory of the Bharatas included India's best copper mines. Even today, the country's largest copper mine is situated at Khetri along the Rajasthan-Haryana border. With the superior bronze and good leadership, the Bharatas were a formidable force. A number of ancient copper items from this period, including weapons, have been discovered in recent decades in southern Haryana, northern Rajasthan and western Uttar Pradesh.

Soon after this great victory, the Bharatas defeated a chieftain called Bheda on the Yamuna. These victories made them the superpower in the subcontinent with an empire that stretched from Punjab, across to Haryana to the area around Delhi-Meerut. Because of their powerful position, their influence would have extended well outside the lands they directly controlled. They must have strengthened their position even further by initiating the compilation of the Vedas. The Rig Veda is full of praise for the Bharata-Trtsu tribe, its chief, Sudasa, and the sage Vashishtha, in a way suggesting that the book was put together under the encouragement of this tribe, probably over several generations following the great battle.

However, the Vedas do not confine themselves to singing the praises of the victors alone; they deliberately include those of sages from other tribes, including some of the defeated ones! The hymns of Sage Vishwamitra, the arch-rival of Vashishtha, are given an important place in the compilation. What does this tell us about the Bharatas? We see a culture that accommodates and assimilates differences rather than impose its own on others. This is a powerful idea and in time, it allowed for people from faraway places like Bengal and Kerala to identify with this ancient Haryanvi tribe.

This is why the Bharatas remain alive in the name by which Indians have called their country since ancient times—Bharat Varsha or the Land of the Bharatas. In time it would come to mean the whole subcontinent. Later texts such as the Puranas would define the land as: 'The country that lies north of the seas and south of the snowy mountains is called Bharatam, there dwell the descendants of Bharata.' It remains the official name of India even today.

Did you know?

In Malay, 'Barat' means west, which is the direction from which Indian merchants came to South East Asia!

After his victories, Sudasa performed the Ashwamedha or horse sacrifice and was declared a Chakravartin or Universal Monarch. The word 'chakravartin' means 'wheels that can go anywhere'—a monarch whose chariot can roll in any direction. The spokes of the wheel symbolize the various cardinal directions. Over the centuries, the symbolism of the wheel would be applied widely. We see it used in imperial Mauryan symbols, in Buddhist art, and even in the modern Indian nation's flag.

Meanwhile, what happened to the defeated tribes? Some of them remained in Punjab, although much weakened. The Druhya tribe was later chased away from Punjab to eastern Afghanistan. Their king Gandhara gave the region its ancient Indian name—still remembered today in the name of the Afghan city of Kandahar. The Puranas also tell us that the Druhyas would later migrate farther north to Central Asia and turn into Mlechhas or foreign barbarians. Nothing more is heard of them. Another tribe called the Purus survived into the Mahabharata epic and probably accounted for King Porus, who fought against Alexander the Great in the fourth century BCE.

Some of the tribes, however, appear to have fled even further after the great battle. Two of them have names that suggest interesting possibilities: the Pakhta and the Parsu. The former are also mentioned by later Greek sources as Pactyians—they could be the ancestors of Pakhtun (or Pashtun) tribes that still live in Afghanistan and north-western Pakistan. Genetically, the Pashtuns are related to Indians and not to Central Asians or Arabs as was previously thought! Similarly, the Parsu are probably related to the Persians because this is the name by which the Assyrians refer to the Persians in their inscriptions.

There is plenty of evidence that links the Rig Vedic Indians to the ancient Persians. The Avesta, the oldest and most sacred text of the Zoroastrians, is written in a language that is almost identical to that of the Rig Veda. The older sections of the Avesta—called the Gathas—are said to have been composed by the prophet Zarathustra himself. They can be read as Rig Vedic Sanskrit by making a minor phonetic change—the 'h' in Avestan is the 's' in Sanskrit. A similar phonetic shift survives in the modern Indian language of Assamese!

The texts are clear that the Avestan people came to Iran from outside. They called themselves the Aryan people. They were aware of the Sapta-Sindhu but not of western Iran, suggesting that they entered the country from outside. Unlike the Vedas, the ancient Persians also talk of an original 'Aryan' homeland and even name the river Helmand in Afghanistan after the Saraswati (i.e. Harahvaiti). Indeed, the Persian identity as 'Aryans' was so strong that their country would come to be known as Land of the Aryans or Iran. As recently as the late twentieth century, the Shah of Iran used the title 'Arya-mehr' or Jewel of the Aryans.

In the Rig Veda, the terms 'deva' and 'asura' apply to different sets of deities and do not denote 'good' or 'bad'. The god Varun, for example, is an asura. However, in later Hinduism, the asuras became demons and the devas became gods. But in the Zoroastrian tradition of Persia, devas refer to demons while the word 'asura' is transformed into Ahura Mazda—the Great Lord! It looks like the devas and the asuras came to be considered as opposites at a later date. What caused this? Did the Parsu have a religious dispute with the Bharatas? As they moved into the Middle East, were the Persians influenced by the Assyrian culture which called their god Assur? We may never know for sure but these are all interesting possibilities.

There is lots of evidence of other Vedic-related tribes in the Middle East in the second millennium BCE. In 1380 BCE, the Hittites signed a treaty with a people called the Mittani. This treaty is solemnized in the name of Vedic gods Indra, Varuna, Mitra and Nasatya. The Mittani appear to have been a military elite who ruled over the Hurrian people living in northern Iraq and Syria. There are records of their dealings with Egyptians, Hittites and the Assyrians. From their names and gods, we can tell that the Mittani were outsiders who had entered the region from the east. Once again, we have evidence to show that the Vedic people moved westward rather than the traditional view that they moved south-eastward into India. The peacocks that recur in Mittani art could be telling us that these people remembered not just the gods but also the animals and birds of the land they had left behind.

The Yezidi people are a tiny religious group of about 1,50,000 people who live today among the Kurds of northern Iraq, eastern Turkey and parts of Armenia. Their religion is an ancient one and they were persecuted for centuries for their faith. Like Hindus, the Yezidis believe in reincarnation and avatars, they pray facing the sun at dawn and dusk, and have a system of endogamous castes. Their temples, which have conical spires, look a lot like Hindu temples and the peacock plays a central role in their religion. But the peacock is not to be found naturally in their lands! The Yezidi themselves believe that they came to the Middle East from India about 4000 years ago, around the time the Harappan Civilization began to disintegrate or perhaps when the Battle of the Ten Kings took place. Does one of these events explain the spread of the R1a1 gene that we discussed in the previous chapter?

The world of the Harappans and the Rig Veda dissolved as the Saraswati dried. No matter what one thinks of the Harappan-Vedic debate, two things are clear. First, geography and the forces of nature played an important role in the evolution of Indian history. Second, the subcontinent has seen a great deal of migration and churn. People, ideas and trade have moved in different directions at different points of time and for different reasons. It is very different from the old view that Indian history is only about one-way invasions from the north-west.

3

Not Just the King of the Jungle

Your parents or grandparents may have told you stories from the Ramayana and the Mahabharata when you were growing up. Some believe that these stories are historical, that they are about real people who actually lived in our world. They claim that it's quite possible that the stories may have been based loosely on real events. Others believe that these stories are fictional and that the characters were not real people but imaginary ones. Whatever the truth about the characters and their history, we can certainly say that the geography described in these epics was based on reality.

From 1300 to 700 BCE, the period that these epics describe, India was going through the next cycle of urbanization. The epics have undergone many changes over the centuries before they reached their current form, so we cannot take all the information from them too literally. While the geography of the Ramayana is along a North-South axis, the Mahabharata is along an East-West axis. The Vedic people gave a lot of importance to the Sapta-Sindhu region but we see a shift in focus in this period.

The North-South axis and the East-West axis described in the epics are along two major trade routes. The Dakshina

UTTARA AND DAKSHINA PATH HIGHWAYS

Modern highways
Uttara and Dakshina path in Gupta and Maurya Eras
Rama's Journey to Lanka

Peshawar
Lahore
Amritsar
NH 1
Chandigarh
New Delhi
Ayodhya
Kanpur
Allahabad
Pataliputra
Nalanda
Chitrakoot
Ujjain
Vidisha
Varanasi
(Sarnath Mughal Sarai)
NH 2
Bharuch
Tamralipti
Kolkata
Nagpur
Panchavati
Ajanta Caves
Daulatabad
Paithan
Hyderabad
NH 7
Kishkindha
Bangalore
Rameshwaram
Kanyakumari

Path (Southern Road) made its way from the Gangetic plains through central India to the southern tip of the peninsula while the Uttara Path (Northern Road) ran from eastern Afghanistan through Punjab and the Gangetic plains to the seaports of Bengal. These two highways have played a very important role in shaping the geographical and political history of India.

Did you know?
The British called the Uttara Path the Grand Trunk Road and Rudyard Kipling described it as 'a river of life as nowhere else exists in this world'. It survives today as National Highway 1 between Delhi and Amritsar and National Highway 2 between Delhi and Kolkata. It is part of the Golden Quadrilateral network.

The Uttara Path was a busy route by the Iron Age. Since then, it has been continuously rebuilt, keeping to the original path for the most part. In contrast, the path of the Southern Road has drifted over time although certain points remained important over long periods. During the early Iron Age, the Dakshina Path probably began near Allahabad where two rivers, the Ganga and the Yamuna, flowed into each other. It then went in a south-westerly direction through Chitrakoot and Panchavati (near Nashik) and eventually to Kishkindha (near Hampi, modern Karnataka). This is the route that Rama is said to have followed during his exile.

Most scholars accept that the Ramayana is the older of the two texts. There are several versions of the epic, including versions that remain popular in other parts of Asia. The one that is most popular and possibly the oldest was composed by Sage Valmiki.

RAMA'S JOURNEY

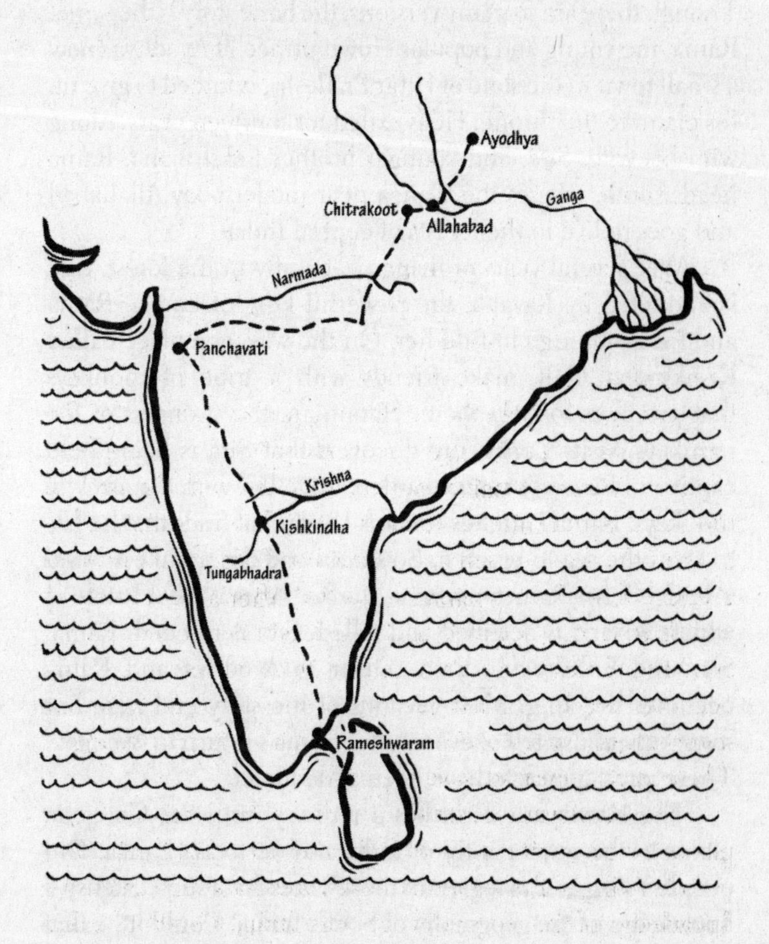

Did you know?
Sage Valmiki was once a bandit. He also belonged to one of so-called lowest castes in Hindu society—what we would today call the Dalits.

Though there are so many versions, the basic story is the same. Rama, the young and popular crown prince of Ayodhya (now a small town in the state of Uttar Pradesh), is forced to give up his claim to the throne. He is exiled for fourteen years. Along with his wife, Sita, and younger brother Lakshmana, Rama heads south, crosses the Ganga near modern-day Allahabad and goes to live in the forests of central India.

After several years of living peacefully in the forest, Sita is abducted by Ravana, the powerful king of Lanka. Rama and his brother go to find her. On the way, at a place called Kishkindha, they make friends with a tribe of monkeys that promises to help them. Hanuman, the strongest of the monkeys, visits Lanka and discovers that Sita is being held captive in Ravana's palace garden. Together with the army of monkeys, Rama marches towards Lanka but finds that he has to cross the sea to reach it. So Rama and the monkeys build a bridge from Rameswaram to Lanka. After a great battle in which Ravana is defeated and killed, Sita is rescued. Rama, Sita and Lakshmana then return to Ayodhya and Rama becomes the king. Most versions of the story end here but some others also tell of events after Rama's return to Ayodhya. These parts appear to have been added later.

The Ramayana describes a journey from the Gangetic plains to the southern tip of India and on to Sri Lanka. Did people living in this region in those times have such extensive knowledge of the geography of South India? Could it be that the names of these places were fitted into the story later? But if one were to visit the sites described in the epic, it is not difficult to believe that Sage Valmiki actually did know about these places. For example, Kishkindha, the kingdom of the monkeys, is a site across the river from the medieval ruins of Vijayanagar at Hampi. This place has strange rock outcrops, caves with Neolithic paintings and bands of monkeys scampering across the boulders.

There are small details in Valmiki's description that ring true even today if one were to look at the landscape. He must have either visited the place himself or heard detailed descriptions of it from merchants travelling the Dakshina Path. For example, the lake of Pampa, surrounded by a ring of rocky hills, where Rama first meets Hanuman, is still a beautiful place with lotuses in bloom and a variety of birds living in it. Not far away from this site is a sloth bear reserve—remember Jambavan, Hanuma's sloth bear friend?

Archaeologists have found the remains of several Neolithic settlements in the area. It is possible that the setting was once home to a Neolithic tribe that used the monkey as a totem. It could be this tribe that is described as the vanaras by Valmiki.

The same can be said of the bridge from Rameswaram to Lanka. There exists a thirty-kilometre-long chain of shoals and sandbanks that links India to the northern tip of Sri Lanka. Are these remains of Rama's bridge or the result of a geological process? Whatever you believe, you will agree that it truly is a remarkable feature! Today, we can see the true scale of the bridge through satellite or aerial photographs but Valmiki, who composed the epic, must have clearly known about its existence for him to write about it.

Ravana is the villain of the Ramayana but he is not shown as a barbarian (Mlechcha). He is portrayed as a learned Brahmin and a worshipper of Shiva. This tells us that the Iron Age Indians considered Ravana and his southern kingdom to be part of the Indian civilization. Even now, the Kanyakubja Brahmins of Vidisha claim Ravana as one of their own and worship him. The exchange of goods and ideas along the Southern Road, therefore, had linked the north and south of India long before its political unification under the Mauryans in the third century BCE.

The Mahabharata is made up of 1,00,000 verses and is said to be the longest composition in the world. Traditionally, it's considered to have been composed by the sage Vyas but it appears to have been expanded over the centuries. We know that a shorter version of the epic was definitely in existence by the fifth century BC but it probably reached its current form centuries later.

More about the Mahabharata

The Mahabharata is the story of the bitter rivalry for control over the kingdom of Hastinapur between the five Pandava brothers and their cousins, the hundred Kauravas. The two first agree to divide the kingdom and the Pandavas build a new capital called Indraprastha. The new capital is so beautiful that the Kauravas are filled with envy. They challenge the Pandavas to a game of dice that is fixed by their maternal uncle, Shakuni. The Pandavas gamble away the kingdom and are exiled for thirteen years. During this time, the Pandavas wander across India. However, when they return after the period of exile, the Kauravas refuse to return the kingdom.

The dispute grows and finally at the great battle of Kurukshetra, in which almost every kingdom of India is said to have taken sides, the Pandavas defeat the Kauravas. Krishna, the leader of the Yadava clan and king of Dwarka, sides with the Pandavas. The last act of the battle takes place away from the main battlefield. Bhima, the strongest of the Pandava brothers, kills Duryodhana, the leader of the Kauravas, in single combat on the banks of the Saraswati. By now, the river would have dwindled to a shadow of its former self. Perhaps no more than a rain-fed seasonal river.

Many of the places mentioned in the Mahabharata are located around Delhi. For example, Gurgaon, which is now full of tall office buildings and shopping malls, was a village that belonged to Dronacharya, the teacher who trained the Pandavas and the Kauravas in martial arts. The name Gurgaon literally means 'village of the teacher'. The Pandava capital of Indraprastha is said to be located under the Purana Quila in Delhi. The site even had a village called Indrapat till the nineteenth century. Excavations between 1954 and 1971 found that there was a major settlement there that dates at least to the fourth century BCE. Pottery shards suggest there may be an older Iron Age settlement nearby.

Similarly, the site of Hastinapur is identified with a site close to modern Meerut. The battlefield of Kurukshetra is nearby, in the state of Haryana. A little further, we have the cities of Mathura and Kashi (Varanasi), which remain sacred places for Hindus even today.

One of the most interesting Mahabharata-related sites is that of Dwarka in the westernmost tip of Gujarat. It is said to have been founded by Krishna as his capital after he led his people from Mathura to Gujarat. Thirty-six years after the Kurukshetra battle, the city is said to have been flooded and taken by the sea. Underwater surveys near the temple-town of Dwarka and the nearby island of Bet Dwarka have come up with stone anchors, a sunken jetty and elaborate walls suggesting the existence of an ancient port in the area. This is yet another example of how nature has directed the course of history!

• MAHABHARATA MAP •

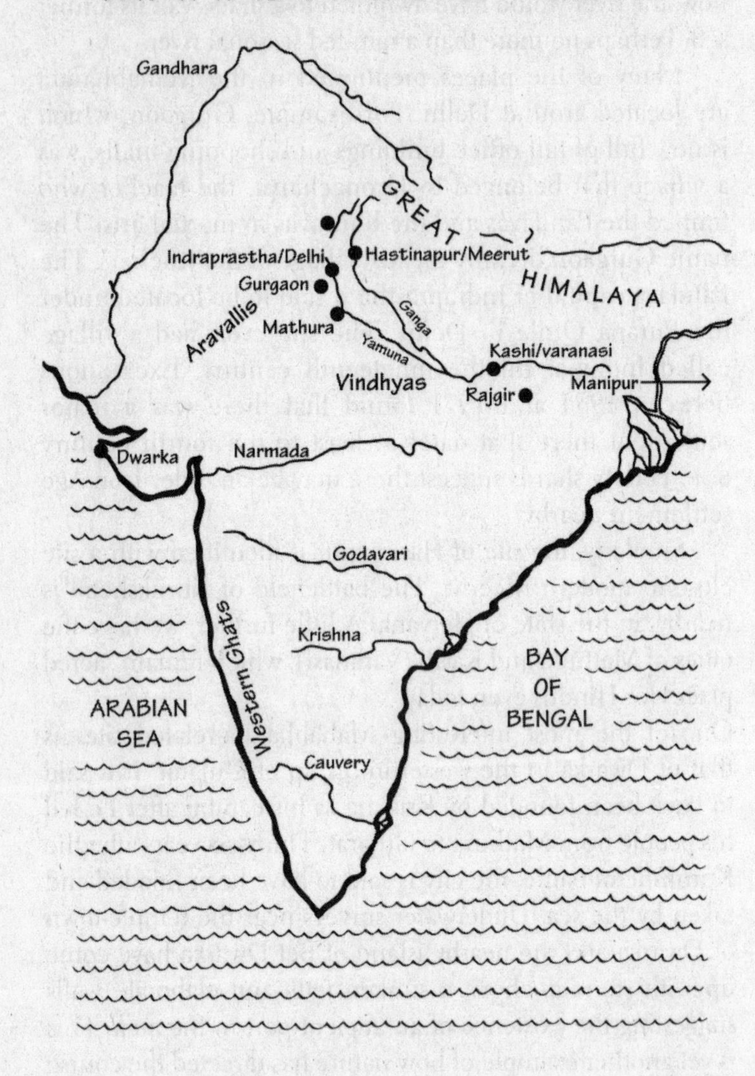

None of this confirms the events of the Mahabharata historically but it strongly suggests that the composers of the epic were talking about real places. Of course, there are gaps between the archaeological findings and the information in the texts, but this is only to be expected after such a long lapse of time.

Did you know?
Till the nineteenth century, the places mentioned in the Greek epic *Iliad* were considered to be mythical. However, excavations have shown that Troy and many of the places mentioned in the epic were actually real! Similarly, Chinese legends about the ancient Shang dynasty (around 1600–1046 BCE) have now been confirmed by modern archaeology.

As mentioned earlier, the Mahabharata largely has an East-West axis. Most of the action takes place around Delhi and the Gangetic plains but the eastern and western extremes of the subcontinent also play an important role. Gandhari, the mother of the Kauravas, is from the kingdom of Gandhara, which is now eastern Afghanistan. It is her cunning brother, Shakuni, who turns the Kauravas against the Pandavas and fixes the game of dice, ultimately causing the war.

On the other geographical extreme, India's North-East is mentioned for the first time in the epic. Arjuna, the most handsome of the Pandava brothers, makes his way to remote Manipur during his years of exile. There he meets the warrior-princess Chitrangada. They fall in love and marry but under the condition that Chitrangada would not have to follow Arjuna back to the Gangetic plains. Their son eventually becomes the king of Manipur and also participates in the Kurukshetra battle.

Did you know?
The people of the tiny Bishnupriya community that still lives in Manipur and neighbouring states trace their origin to the Mahabharata. They speak a language that is related to Assamese and contains Tibetan-Burman words but still preserves several features of the archaic Prakrit!

Since the Kurukshetra battle is said to have involved all the tribes and kingdoms of India, the Mahabharata gives us long lists of kingdoms, clans and cities. Many of them were probably added to the text in later times. However, these lists give us a rough idea of the Indian world view during the Iron Age. The name Mahabharata itself is interesting as it can be read to mean 'Greater India'. This would make sense for an epic that claims to tell a story involving all the clans of the subcontinent. Also, the name makes a reference to Emperor Bharata who is said to have conquered the whole country (but he plays no important role in the central plot). The epic is told as a history of the Bharata people. Since there is no evidence of the all-conquering Emperor Bharata, it is possible that this is an echo of the powerful Bharata tribe mentioned in the Rig Veda. Did Sudasa's victory against the ten tribes create a dream of nationhood as we understand it today?

Let's look at India's neighbour China and its ideas of nationhood. Long before the country was united into an empire by Qin Shi Huangdi in the third century BCE, there was a strongly held belief that the country had once been united under the 'Yellow Emperor' and his four successors. There is no archaeological evidence to support such a grand empire but it has been a very powerful idea throughout Chinese history.

The notion of nationhood is not a simple one. It has meant different things to different people at different points in time. The Partition of India in 1947, for instance, was partly due to a fundamental difference in views about the nature of India's nationhood. Still, it is important to understand how Bronze Age ideas, shaped in the Iron Age, have influenced the way people have viewed themselves since.

The epics also suggest a shift of political power to the eastern Gangetic plains during the Iron Age. It is more obvious in the Ramayana as the kingdom of Ayodhya is in the east. In the Mahabharata, most of the action takes place near Delhi in the north-west but, even here, we are told of the powerful kingdom ruled by Jarasandha in Magadh (modern Bihar). Even Krishna was forced to shift his people from Mathura to Gujarat because of the repeated raids of the Magadhan army. The rise of Magadh would play an important role later in Indian history—we'll read more about that later.

Why was Magadh so successful? It could be because of its geographical access to three important resources—rice, trade and iron. The kingdom not only had control over very fertile lands but was also served by a number of rivers including the Ganga itself. Moreover, the kingdom controlled the trade between the Uttara Path between the North-West and the emerging seaports of Bengal. It also had access to iron ore from what is now Jharkhand. The kingdom's first capital, Rajgir (also referred to as Rajagriha or 'king's home'), is defended by the hills and sits strategically between the fertile farmlands to the north and the mines to the south. In short, Magadh was able to feed large armies and arm them with iron weapons.

Enter the Lion

India is the only country in the world where lions and tigers co-exist. As mentioned in Chapter 1, tigers evolved in East Asia and probably entered the subcontinent around 12,000 years ago. Soon, they spread across the subcontinent. Tigers commonly appear in Harappan art and seals. But the lion is nowhere to be seen! None of the main Harappan sites have thrown up any images of the lion. This is very strange because we see the lion being given a lot of importance in later Indian culture.

The tiger hunts at night in the dense jungle. It's an object of fear. The lion, on the other hand, hunts in the open. With its shaggy mane and confidence, it's usually seen as a symbol of power. Every culture that has encountered the lion has tended to give the animal a special status. Even countries that have never had lions, such as Britain and China, have used the animal to symbolize power. In ancient Mesopotamia in the second millennium BCE, only the king could hunt lions. In ancient Egypt as well, only royalty could go lion hunting. Amenhotep III (1391–52 BCE) killed as many as 102 lions in the first decade of his rule! At Beital-Wali in Lower Nubia, a tame lioness is shown near the throne of Rameses II (1290–24 BCE) with the inscription 'Slayer of his Enemies'. Five centuries later, the court records of the Assyrian king Ashurabanipal II (884–859 BCE) mention that he killed 370 lions with his spears!

The lion is also represented in many sculptures, friezes and paintings in Ancient Egypt and Mesopotamia. The Sumerian goddess Nana, the Assyrian goddess Ishtar and the Persian goddess Anahita are all associated with the lion and sometimes depicted riding the lion—rather like the Hindu goddess Durga.

The lion then was an important animal in art, culture, royal symbolism and religion in the Middle East from a very early period. So why did the Harappans ignore such a glorious beast?

This was probably because the animal was not common in the subcontinent till after the collapse of the Harappan Civilization. Before 2000 BCE, north-west India was much wetter than it is today with higher rainfall. The Saraswati river would have been in full flow. The lion hunts in open grasslands. It could not have gone to the dense jungles, which were full of

Goddess Durga on the mighty lion

tigers. But when the climate became drier and the Saraswati began to dwindle, there would have been a phase when the lions from Iran could have made their way through Baluchistan. This is possibly why the earliest artefact depicting a lion in the subcontinent, a golden goblet, was found in Baluchistan.

As Harappan urban centres were abandoned and populations migrated to the Gangetic plains, the lions would have taken over the wilderness. Over time, they would travel as far east as Bihar and north-western Orissa, living in many places along with tigers. They did not make further inroads into Eastern and Southern India as the forests were too thick and the climate too wet. The Rig Veda does mention the lion but it does not give it as much importance as it does the horse or the bull. How did the Vedic people know of the animal if it did not exist in the Sapta-Sindhu heartland? Could it be that the word for lion, Singha, was simply something that they used to describe all big cats, including tigers? The dual use of the word is responsible for the naming of Singapore! More on that later. Some experts feel that the Vedic description of a hunt suggests lions rather than tigers. It's possible that while the lion was not common in the Sapta-Sindhu region, the Vedic people may have encountered it in lands to the west of the Indus (remember the lion goblet found in Baluchistan?). However, we don't know enough about this period to be absolutely sure.

Whichever way the lion entered the subcontinent, it quickly became a cultural symbol in the land, just as it had been in the Middle East. The word for 'throne' in Sanskrit is 'singhasana', which means 'seat of the lion'. We've already seen how Durga, the Hindu goddess of strength and war, is shown riding a lion while slaying a demon. The Mahabharata repeatedly uses the image of a lion to convey strength and vigour. Even now, communities that are proud of their martial tradition, like the Rajputs and the Sikhs, commonly use Singh as their surname. Yes, Singh means lion!

The lion also plays a significant role in the *Mahavamsa*, a Pali epic from Sri Lanka. According to it, the Sinhalese people are the descendants of Prince Vijaya and his followers who sailed down to Sri Lanka in the sixth century BCE from what is now Orissa and West Bengal. Prince Vijaya was the son of a lion and a human princess—and this is why the Sinhalese call themselves by this name, which means 'lion people'. The country's national flag has a lion holding a sword.

You may have heard of the LTTE, a separatist organization in Sri Lanka fighting for *eelam* or a land of their own for the Tamils. The Tamil rebels chose to call themselves the 'Tigers', as opposed to the Sinhalese who use the lion as their symbol. The ancient rivalry between the two big cats is still dominant in our imagination though both animals are close to extinction.

There are now a mere 411 Asiatic lions left in the wild. The Gir National Park in Gujarat is their last refuge. Less than 200 years ago, this magnificent animal could be found around Delhi and was probably common in the Aravalli ridges in the south of Gurgaon. Now these places are full of highways and speeding vehicles. The lion was last seen in Iran in 1942 and in Iraq in 1917.

Chopped Nose? No Problem!

By the late Iron Age (eighth to fifth century BCE), a number of urban centres were growing to the size of the old Harappan cities. Kausambi, near today's Allahabad, is said to have been founded after the king of Hastinapur, a descendant of the Pandavas, who was forced by a devastating flood to shift his capital further east. Spread over an area of 150–200 hectares, Kausambi had a population of around 36,000 people at its peak. There were other major cities like Rajgir and Sravasti that were equally large. These were similar to Mohenjodaro, the largest of the Harappan sites, which had a population of around 40,000. It's hard to say just how many people were there in the entire subcontinent but it's likely to have been around 30 million.

The late Iron Age towns were fortified with moats and ramparts. Wood and mud bricks were materials commonly used to build but the people had not forgotten the kiln-fired bricks that the Harappans used. Kausambi has many buildings that used this technology. The towns also had drains, soakage pits and other urban facilities though the designs were different from those of the Harappans. But the courtyard continued to exist and the streets were levelled to allow the movement of vehicles with wheels.

The Ganga would have been full of merchant boats travelling between Kausambi, Kashi and Pataliputra (modern

Patna). There would have also been ships that could travel across the ocean. The legend of Prince Vijaya who travelled from India to Sri Lanka suggests coastal trade links along the Bay of Bengal, extending from Bengal to Sri Lanka. Both the Uttara Path and the Dakshina Path would have been busy highways with people carrying goods and ideas. This was a time of great intellectual expansion—the philosophies of the Upanishads, Mahavira and Gautama Buddha are all from this time period!

The Buddha was born in Kapilavastu (on the Indo-Nepal border) but he attained enlightenment at Bodh Gaya, just south of the old Magadhan capital of Rajgir. But he did not deliver his first sermon in Bodh Gaya, the nearby towns and villages or even in the royal capital. Instead, he headed west to Varanasi (also called Kashi). Why did he go all the way there to spread his message?

This may have been because Varanasi stood at the crossroads between the Uttara Path and a highway that came down from the Himalayas and then continued south as the Dakshina Path. It was already a place where goods and ideas were being exchanged.

Did you know?
Even today, the east-west National Highway 2 meets the north-south National Highway 7 at Varanasi. NH7 then runs all the way down to the southern tip of India. The alignment of the modern north-south highway runs somewhat east of the ancient trade route but isn't it amazing that the logic of India's transport system has remained the same? Even when the British built the railways in the nineteenth century, they used Mughalsarai—just outside Varanasi—as the nerve centre of the railway network.

When the Buddha went there in the sixth century BCE, Varanasi was already a big urban settlement built on the Ganga. The city was built between where the Varuna and Asi streams flow into the Ganga and was therefore called Varanasi. The Varuna still exists but the Asi has been reduced to a polluted municipal drain.

And so, the Buddha chose to deliver his first sermon in a deer park at Sarnath, just outside Varanasi, the centre for commercial and intellectual activity. The spot is not sacred for the Buddhists alone. Just outside the site, there is a large Jain temple dedicated to the eleventh 'tirthankara'. Similarly, the archaeological museum next door contains many idols and artefacts of the Brahminical tradition. The name Sarnath is a short form for Saranganath, meaning Lord of the Deer, another name for Shiva. Varanasi has always been a very important place for the Hindus, especially those who worship Shiva. It may explain why the Buddha found a park with sacred deer at this place. There is a temple dedicated to Saranganath, less than a kilometre from the archaeological site, even now.

Apart from religious philosophy, the period also saw the systemization of Ayurveda, India's traditional medical system. Sushruta, who lived near Varanasi, put together the medical knowledge of that time and also included a long list of sophisticated surgical instruments and procedures. There are detailed descriptions of plastic surgery, surgeries on the eyes, and other complex procedures. There is even information on the dissection of dead bodies to learn about anatomy.

Medieval India refers to the Post Classical Era, i.e. eighth to eighteenth century CE in the Indian subcontinent. It is divided into two periods: the 'early medieval period', which lasted from the eighth to the thirteenth century, and the 'late medieval period', which lasted from the thirteenth to the eighteenth century in some definitions, though many end the period with the start of the Mughal Empire in 1526.

However, much of this knowledge was lost in the medieval era, we don't know why. Possibly the destruction of centres of learning during the Turkic invasions is partly responsible for this. Still, some techniques survived and were witnessed by European visitors in the eighteenth century. This includes the famous 'rhinoplasty' operation that took place in Pune in March 1793, which greatly influenced plastic surgery in Europe and the rest of the world. Cowasjee was a Maratha (more likely Parsi) bullock-cart driver with the English army during its campaigns against Tipu Sultan of Mysore. He was captured and had his nose and one of his hands cut off. After a year without a nose, he and four others who had suffered a similar fate allowed an Indian surgeon to use the skin from their foreheads to repair the noses. We know little about the surgeon but two senior British surgeons from Bombay Presidency witnessed this operation and sent back detailed descriptions and diagrams. The publication in Europe in 1816 of their account gave birth to modern plastic surgery.

Of course, there was cultural and intellectual activity of this period happening in other parts of the subcontinent too and not just in the Gangetic heartland. For example, Panini, the famous grammarian who standardized the Sanskrit language in the fifth century BCE, was said to have been born in Gandhara (eastern Afghanistan) and lived in Taxila (near modern Islamabad). This part of the subcontinent was about to see the first attempt by a European power to conquer India.

He Came, He Saw, He roared

The world of small tribal kingdoms described above went through a major change in the third and fourth centuries BCE. This happened around the same time all over the world. This was not really because of a change in technology but because of a change in political ideas and ambition. Within a couple of generations, quite a few leaders were inspired by the idea of an empire. These leaders then began looking at how they could conquer other parts of the world.

The first of the empire-builders was Cyrus the Great of Persia in the sixth century BCE. But it is only in the fourth century BCE that we begin to see empire-building on a totally different scale. In China, King Hui of Qin began a cycle of conquest around 330 BCE that would lead to building the first empire under Shi Huangdi a century later. At around the same time, Alexander the Great took control of Greece, Egypt, the Levant, Mesopotamia, Bactria and Persia.

Then, in the winter of 327–326 BCE, Alexander marched into India. He built an alliance with Ambhi, the king of Taxila. Together they defeated Porus on the banks of the Jhelum. It's possible that the name Porus refers to the Puru tribe that had inhabited the area since Rig Vedic times. Alexander wanted to march eastwards but his troops were tired. There were also stories about a large Magadhan army waiting to attack them in the Gangetic plains. With an unwilling army, Alexander had little choice. He decided to return. But he did not go back the way he had come—he chose to sail down the Indus under the mistaken belief that the Indus was part of the upper reaches of the Nile. He thought that if they just sailed down the Indus, they would end up in the Mediterranean!

Alexander and his people reached this conclusion because of the similarities between the plants and animals of India and those of the upper reaches of the Nile. As they sailed down the Indus, they defeated many tribes and destroyed several settlements. There is also a fascinating account of how a local chieftain entertained Alexander with a gladiatorial match between a lion and a group of ferocious dogs that he claimed had been bred from tigresses!

On reaching the sea, Alexander discovered his mistake. They were then forced to march along the dry Makran and Persian Gulf coast—the same route that early humans had used when they migrated east to the subcontinent. However, climatic conditions and the coastline had changed a lot since then. Without proper maps, provisions and water, the desert proved to be a nightmare. Soldiers and pack animals died in large numbers. Much of the wealth they had acquired from their conquests had to be abandoned because there weren't enough men and animals to carry the loot. When Alexander's army finally reached Babylon they remained undefeated but they had suffered heavy losses. Alexander died soon afterwards, possibly poisoned by followers who no longer believed in his leadership. His empire was divided up amongst his generals and his young son was murdered. It was the lack of geographical knowledge that proved to be Alexander's undoing, not a sword. As we shall see, when Europeans attempted to take control of India two millennia later, they would take great care to map it.

Alexander's invasion is not mentioned directly in Indian texts but Greek writers have left us detailed accounts of their adventures. Some of them seem quite fantastic—there's one about giant ants that were used to dig for gold! But for the most part, their observations were accurate. Nearchus, Alexander's admiral, tells us that Indians wore clothes made from white cotton. Their lower garment reached below the knee, halfway to the ankles. The upper garment was thrown over the shoulder and the turban was worn on the head.

Nearchus was describing the dhoti and angavastra—clothes that have been worn since Vedic times and continue to be used even today! He goes on to say that wealthy Indians

Fashion in the Iron Age

flaunted ivory earrings and carried umbrellas against the sun. They also wore thick leather sandals with elaborate trimmings and thick soles to make themselves look taller!

Alexander's invasion did not really have much of an impact in the Indian heartland but it did trigger a chain of events. One that would lead to the founding of India's first great empire—that of the Mauryans. The empire was created by two extraordinary characters: Chanakya (also called Kautilya) and his student Chandragupta Maurya. Most empires were created by princes and warriors but Chanakya was a professor of Political Economy in Taxila. When Alexander entered into an alliance with the king of Taxila, the Brahmins of the city opposed this. Historical accounts say that Alexander had several of them hanged to death.

According to legend, Chanakya travelled east to Pataliputra (modern Patna), the capital of the powerful kingdom of Magadh to ask for help against Alexander. But he was insulted and thrown out. An angry Chanakya decided to return to Taxila to plot his revenge. On the way he came across a boy called Chandragupta Maurya. There are many stories about Chandragupta's origins and how the two met, but these cannot be verified.

Chanakya took the boy back with him and began to train him to become a king. He also wrote the *Arthashastra* (Treatise on Prosperity), a detailed manual on how to run the future empire. When Alexander died, Chanakya decided that this was the right time to put together a band of rebels and fight for power. However, their initial efforts at throwing over the Nanda king of Magadh failed. It is said that Chandragupta had to flee into the forests to escape. He was so tired that he fell into a deep slumber. A lion appeared and licked him clean of all the grime and dust. Then it stood guard over him till he awoke. When

Chandragupta realized what had happened, he accepted it as a good omen and attacked the Nandas once again. It's quite possible that this rather fantastical tale was cooked up by later Mauryan supporters but once again, it underlines the symbolic importance of the lion.

After many years of effort, Chanakya managed to put together a large army, possibly with the help of the hill tribes of Himachal. He and Chandragupta slowly took control of the north-west of the country. Then they set their eyes on the Gangetic plains. Around 321 BCE, they defeated the Nanda king of Magadh and became the most powerful in the subcontinent. However, Chanakya did not take the throne for himself. He crowned Chandragupta Maurya instead. Then they spent over a decade establishing control over central India.

By around 305 BCE, Chandragupta felt confident enough to directly confront the Macedonians left behind by Alexander. One of Alexander's most trusted generals, Seleucus Nikator, was in control of the conqueror's Asian domains, including Persia and Central Asia. He also laid claim to the Indian territories conquered by Alexander. Judging by the terms of a treaty between the two in 303 BCE, it appears as if the Mauryan army decisively won the war. Chandragupta gained control over Baluchistan and Afghanistan. Seleucus also gave his daughter in marriage to a Mauryan prince, possibly Chandragupta himself or his son.

For three generations, the Mauryan Empire covered the whole subcontinent from the edge of eastern Iran to what is now Bangladesh. Only the southernmost tip of India was out of their direct control. At its height, it was the largest and most populous empire in the world, much greater than Alexander's domains and those of Shi Huangdi in China. It also lasted for a much longer duration as a complete unit.

But there was something unique about this empire-building. Chanakya was happy to remain a minister and according to one version, he actually went back to teach in Taxila once the empire had been stabilized. Chandragupta Maurya himself placed his son Bindusara on the throne and became a Jain monk, giving up all his wealth and comforts. He took the Dakshina Path and travelled down to Sravana Belagola (in Karnataka) and according to Jain tradition, starved himself to death to cleanse his soul. The hill on which he spent his last days meditating and fasting is still called Chandragiri in his honour.

The idea of renouncing power has remained a powerful theme in later Indian history. When India became independent in 1947, Mahatma Gandhi refused all positions of power and made way for his protégé Jawaharlal Nehru to become modern India's first prime minister.

The second Mauryan emperor, Bindusara, ruled from 297 to 272 BCE. His reign was mostly peaceful. There are records that talk of how the Mauryan emperor exchanged ambassadors and improved trade relations with Alexander's successors in the Middle East. There is also a tale that Bindusara asked Antiochus of Syria to send him figs, wine, and a Greek scholar. Antiochus sent him the figs and wine but refused to send the scholar, saying that Greek law did not permit the sale of scholars!

There seems to have been a struggle to decide Bindusara's successor. The winner of this clash was Ashoka who was crowned in 268 BCE. He was not his father's chosen successor but he ruled the empire for forty years. In 260 BCE, Ashoka expanded the empire for one last time to include

Kalinga (roughly modern Orissa). He now ruled the whole subcontinent except for the small kingdoms of the extreme south with whom he had friendly relations.

These southern kingdoms were called Chola, Pandya, Keralaputra and Satiyaputra. The Cholas would remain a powerful clan for the next one and half millennia and head a powerful empire of their own in the tenth and eleventh century CE. We will read more about them later. Keralaputra, if you haven't guessed already, lent its name to the state at the south-western tip of the Indian peninsula—Kerala.

From Pillar to Pillar

Ashoka is the first Indian monarch who has left us artefacts that belong to his reign without any doubt. The name Ashoka does not appear on any major declaration or inscription. They were issued by a king who called himself 'Piyadassi' or 'Beloved of the Gods'. However, there is strong evidence that links Piyadassi to Buddhist legends about a great king called Ashoka. This evidence suggests that Ashoka was Chandragupta Maurya's grandson.

The best-known Ashokan artefacts are a series of inscriptions engraved on rocks and stone pillars scattered across the empire. These pillars and inscriptions have been found across the subcontinent from Afghanistan in the north to Karnataka in the south, Gujarat in the west to Bengal in the east. They are also scattered across the northern plains, including one in Delhi (near Greater Kailash). We can assume that there must have been many more pillars and inscriptions that did not survive over the centuries. Still, what remains is impressive and gives us a sense of the scale and extent of the Mauryan Empire.

These artefacts have been of great interest since they were interpreted in the nineteenth century. This is not surprising, given their age as well as their contents! Ashoka openly regretted the invasion of Kalinga and the bloodshed it caused. He said, 'On conquering Kalinga, the Beloved of the Gods felt remorse, for when an independent country is conquered, the slaughter, death and deportation of the people is extremely grievous to the Beloved of Gods and weighs heavily on his mind.' He asks his subjects to be good citizens while committing himself to their welfare.

The Kalinga campaign was brutal. About 1,50,000 people were forced to move away from their homes, over 1,00,000 were killed, and even larger numbers eventually died because of wounds and famine. India's population at this stage would have been around sixty-five million. So many deaths at this stage would have been devastating for a small province like Kalinga. Excavations at Kalinga's capital of Tosali reveal structures that still bear marks of this attack. The large number of arrowheads found embedded in a small section of the ramparts tell of a blizzard of arrows.

Ashoka appears to have regretted his decision because of the suffering it caused. Very unusual for any emperor from any era, especially if you were to contrast it with the brutal rule of the First Emperor of China at about the same time. However, at the end of the day, Ashoka was a politician. And one must take a politician's statements with a pinch of salt! These inscriptions are what Ashoka wanted people to remember of him. While he expresses his regret, notice how he did not at the same time offer to free Kalinga and its inhabitants!

Though these inscriptions are very interesting, historians have focused too much on the noble sentiments expressed in them rather than on the pillars themselves. Around 40–50 feet high, the stone columns are impressive structures often capped by a lion or lions. This is an animal that has been associated with the Mauryans since Chandragupta's time. In some of the pillars, the lions are accompanied by the chakra or wheel. Historians often associate this with the Buddhist 'dharma-chakra' but it is possible that this symbolizes the Chakravartin or Universal Monarch. The pillars and the lions are a clear expression of imperial power. They were the Mauryan way of marking territory.

The Mauryan Lions of Sarnath

Ashoka's average subject would have been illiterate and unable to read the inscriptions. But just the sheer might of those pillars would have left no doubt in their mind about the power of their emperor. The use of such structures to signal power is not unique to the Mauryans or even to India. The ancient Egyptians and the Romans also used them. In India, the successors of the Mauryans raised their own columns and also inserted their own inscriptions on the Ashokan ones.

The Mauryan lions and pillars were mostly made from sandstone quarried at Chunar, near Varanasi, where the

Ganga nudges the Vindhya range. We now know the exact location of the quarries to the south-west of Chunar fort, close to the famous Durga temple. Stone is still quarried here, and one can see some of the ancient quarries as well as cylindrical blocks of unfinished stone abandoned by the ancient stone-cutters. Some of them bear inscriptions that tell us when the stone was originally quarried.

The Mauryans rolled the stones to the river and then transported them by boat to workshops near Varanasi, just as the ancient Egyptians transported stone blocks down the Nile to construct their temples and pyramids. Though various irrigation projects these days have drastically reduced the water-flow in the Ganga, it is still possible to make the journey by boat from Chunar to Varanasi.

Archaeologists have found remains of workshops along the river where this stone was carved and polished. The stone used to carve the Sarnath lions, modern India's national symbol, would have made this journey from quarry to workshop and then to Sarnath. There are still several stone-carvers who work on Chunar sandstone in and around Varanasi. What's more, some of them are still carving lions to adorn homes and temples and the new sculptures all bear the same 'grin' that one sees on the Mauryan lions!

Later rulers understood the symbolic meaning of the Mauryan columns and were always keen to make them their own in some way. This is why the emperors of the Gupta and Mughal dynasties went out of their way to put their own inscriptions next to those of Ashoka. Feroze Shah Tughlaq, the fourteenth-century sultan of Delhi, even had two of the pillars shipped to his newly built palace complex! It was therefore not very surprising that when India became independent, Mauryan lions and the chakra became the country's national symbols. These symbols have always stood for the power of the State, after all.

Some say that Ashoka himself took advantage of the symbolism that already existed. After all, there are legends that associated Chandragupta Maurya, Ashoka's grandfather, with lions. Some scholars even argue that a few of the Ashokan columns may actually have been put up by his predecessors and that Ashoka merely added his inscriptions to them.

Ashoka ruled till he died at the age of seventy-two in 232 BCE. The Mauryan Empire collapsed soon after. Why did the empire collapse so quickly after Ashoka? Some feel that this was because of Ashoka's interest in Buddhist philosophy. They say this must have sapped the morale of the army and the administration. We're not sure what exactly happened but there is evidence to suggest that the empire had already begun to crumble in Ashoka's later years. There are many stories about fights and struggles within the royal family which made the ageing emperor powerless.

It's also possible that the real problem was that Ashoka held on to power for too long. Though he was keen to follow the path of righteousness, he found it difficult to give up his power even when he could no longer rule effectively. Contrast this with the attitude of Chanakya and his own grandfather, the founders of the empire. The problem of ageing rulers clinging on to authority is something that has haunted India through the centuries.

Their Way and the Highway

By the time the Mauryan empire was established, the second cycle of India's urbanization had been underway for a millennium. Taxila in the north-west was not just a vibrant city but an important intellectual hub. In the east, Tamralipti was established as a major port; it is likely that Emperor Ashoka sent his son Mahindra on a mission to Sri Lanka from there. The site is located across the river from Kolkata and is not far from the port of Haldia.

Did you know?
The name 'Tamralipti' means 'full of copper' and may have originally been linked to the export of copper goods. Excavations have revealed punch-marked coins from this period.

The capital Pataliputra was the most important city in the empire. Megasthenes, the Macedonian ambassador to Chandragupta, tells us that Pataliputra was surrounded by massive wooden fences with sixty-four gates and 570 watchtowers. The city was shaped like a parallelogram 14.5 km in length and 2.5 km in breadth. Even if one does not take the numbers literally, it still suggests a very large city. Tower-bases and stockades found from excavations support this.

The main gates had wide wood-floored walkways with bridges across a moat system. The moat system, fed by the Son river, was almost 200 metres wide on the landward side. Along the Ganga, wooden piles were sunk into the mud to protect against inundation. Brick and stone were used to construct buildings inside the walls, especially for important structures. However, wood was a common building material and fires could cause a lot of damage. Megasthenes tells us that he had visited all the great cities of the east but that Pataliputra was the greatest city in the world.

What was it like to live in a Mauryan city? Chanakya's *Arthashastra* has a long list of municipal laws that gives us a good insight into the civic concerns of that time. For example, there were traffic rules that said that bullock carts were not allowed to move without a driver! A child could only drive a cart if accompanied by an adult. Reckless driving was punished except when the nose-string of the bullock broke accidently or if the animal had panicked.

The *Arthashastra* also contains instructions on how to dispose of waste, and rules for buildings, maintaining public spaces like parks and even against encroachment into a neighbour's property. Chanakya didn't approve of nosy neighbours—there's a rule against interfering in the affairs of a neighbour! There were also specific rules against urinating and defecating in public spaces. Fines were imposed on those who committed such offences near a water reservoir, a temple or a royal palace.

Obviously, all of this indicates that this was a society that had a sophisticated understanding of urban life. Was all this relearned in the Iron Age or was it passed on from the Harappan way of life?

Most people in the Mauryan Empire lived in villages, and Chanakya attached a great deal of emphasis to agriculture, animal husbandry and land revenue. He gives detailed instructions on the management of forests, especially on elephants. Summer was apparently a good time to catch elephants and twenty-year-olds were considered to be of the ideal age. At the same time, the capture of pregnant or suckling females and cubs was strictly forbidden.

The establishment of the Mauryan Empire created a stable environment that encouraged trade within and outside the subcontinent. There were major imperial highways crossing the country. The most important of these extended from Taxila to the port of Tamralipti in Bengal. The Mauryans were merely formalizing the Uttara Path that had already existed for over a thousand years. Megasthenes probably used it to visit Pataliputra which he praised so much.

The Dakshina Path also remained an important highway, especially because of the extensive Mauryan conquests in the south. However, the course of the road had shifted somewhat eastwards since the Iron Age. The new route passed through Vidisha and then made its way to Pratishthana (modern Paithan in Aurangabad, Maharashtra). It is possible that by the Mauryan period, a branch of the southern highway already connected Ujjain to the ports of Gujarat. This route would become more important during the Gupta period.

Meanwhile, the sea routes were becoming more and more important. We know that by Mauryan times, there was coastal shipping between Tamralipti in Bengal and Sri Lanka. Links with South East Asia were also being established. It is likely that the ships initially hugged the coast, but as we shall discuss in the next chapter, nautical skills and shipbuilding technology were soon advanced enough to allow merchants to directly cross the Arabian Sea and the Bay of Bengal.

Message to the Future

By the time the Mauryans created their empire, Indian civilization was already well-developed. It was also interested in recording its own history, going by the long list of kings preserved in the Puranas and elsewhere. These records may or may not be perfectly accurate but they show that people in these times wanted future generations to know about their land and its rulers. The Mauryans drew inspiration from these practices, which already existed before their arrival, including the idea of Chakravartin or Universal Monarch. However, they introduced an important innovation—the use of columns and rock inscriptions to record their presence. Like monarchs around the world, Ashoka wanted to be remembered. These structures were not only to mark territory and impress subjects but also to speak to us, the future generation.

Later rulers, who understood what the Mauryans were trying to do, created their own monuments and also tried to link themselves to the Mauryans. They continued to do this centuries after the Brahmi script had been forgotten and these original inscriptions could no longer be read. Even rulers of foreign origin did not break this chain.

In Girnar Hill in Junagarh, Gujarat, there is a rock outcrop at the foot of the hill with an Ashokan inscription. More than three centuries later, a Saka (i.e. Scythian) king called Rudradaman added his own inscription next to it. This second inscription records the restoration of the Sudharshana reservoir. The reservoir was originally constructed by Pushyagupta, Chandragupta Maurya's provincial governor. It was completed during Ashoka's time by Tushaspa, an official of possibly Greek origin. The inscription goes on to say that the reservoir was severely damaged by a great storm and floods in the year 72 (probably 150 CE). This was considered a catastrophe by the local people but Rudradaman proudly says that he had the lake restored within a short time, without resorting to forced labour or extra taxes.

Another 300 years later, the Sudharshana lake burst from its banks again. There is a third inscription on the rock that tells us that this time it was repaired by Emperor Skandagupta of the Gupta dynasty in 455–56 CE. If this is not a sense of history, what is?

Girnar is remarkable not just for this reason. If you climb up the hillock behind the rock inscriptions, above the Kali shrine, you will see Girnar Hill with all its ancient Hindu-Jain temples on one side. On the other side is the Junagarh fort and town. The fort is one of the oldest in the world and according to legend, built by Krishna's army. The very name Junagarh means 'old fort'. Over the centuries, Saka, Rajput, and Muslim kings ruled over it. As we shall see, Junagarh would be the focus of important events when India gained independence in 1947. Barely half an hour's drive away is Gir National Park, the last home of the Asiatic lion.

4

Dip Dip Dip, It's a Stitched Ship!

Once the Mauryan rule had collapsed, the outer edges of the empire soon broke into smaller kingdoms. A part of the empire continued under the Shunga dynasty.

Even though the empire had broken down, it was still a big area and the northern and southern trade routes continued to be busy. The royal court maintained international diplomatic relations; there's a stone pillar raised by Heliodorus, the Greek ambassador in Vidisha, a major pit stop on the Dakshina Path.

The north-western parts of the subcontinent came to be occupied by Indo-Greek kingdoms that evolved a culture based on a mix of Indian, Greek and Bactrian elements. But once again, nature would play a role in the course of history.

In the first century BCE, there was severe snow in the area we now call Mongolia. This led to a famine. A fierce tribe of nomads called the Xiongnu lived in this area. We are not sure who exactly these people were but the Mongols are probably descended from them. These tribes had created a lot of trouble for early Chinese civilizations and it was because of them that the First Emperor decided to build the earliest version of the Great Wall of China.

Because of this famine, the Xiongnu migrated into the lands of another Central Asian tribe called the Yueh-Chih. The Yueh-Chih had to move out and they, in turn, forced the Sakas (Scythians), the Bactrians and Parthians to vacate their lands. One by one, these groups were forced to move into the subcontinent. Thus, Afghanistan and North West India saw a succession of invasions and migrations. For several centuries, this region continued to remain unstable.

Despite this instability, there were also some peaceful periods when trade and culture grew. Taxila remained a centre of learning and new urban centres appeared, especially under Kushan rule. Buddhist ideas made their way into Central Asia and then eventually to China. As we saw, the heart of Indian civilization had already shifted from the Sapta-Sindhu region to the Gangetic plains during the Iron Age. Now, the action moved to the coasts due to a boom in overseas trade.

Overseas trade was not new to India. As we have seen, the Harappans traded actively with Mesopotamia. In the Iron Age, centres like Dwarka may have continued to trade with these places. By the time of the Mauryans, Tamralipti was a busy port with links as far as Sri Lanka. We also know that the empire had diplomatic and trade interactions with the Greek kingdoms of the Middle East. However, it was from the second century BCE that trade with the Greeks, Romans and South East Asia really grew in volume.

A Tamil epic from this period—*Silapaddikaram*—tells us about the story of two lovers—Kannagi, daughter of a captain, and a merchant's son named Kovalan. The epic describes the busy port of Puhar (or Kaveripatnam) as a place that was envied by great kings for its immense wealth brought in by the merchants.

The literature from this period talks quite a lot about trade. This is especially true of the Sangam anthologies. These collections of early Tamil poetry seem to have been put together in a series of conferences which probably took place between the third century BCE and the sixth century CE. Madurai seems to have been the venue for most of these gatherings. Some say that the tradition began even earlier in another city, also called Madurai, which was built along the coast. Apparently, this city, too, like Dwarka, was swallowed by the sea.

Unfortunately, many scholars studying Sangam literature only try to prove the 'purity' of Dravidian culture. They wish to show that it had nothing to do with the 'Aryan' influences from the north. This is quite ridiculous. First, the society described in the poems is full of trade and exchange with the rest of India as well as foreign lands. It is a world that is open to absorbing influences from everywhere and is actually welcoming of this. Secondly, the Sangam poets clearly had strong cultural and religious connections with the rest of the country. They knew of Buddhist, Brahminical and Jain traditions that are of 'northern' origin. Even when local gods like Murugan (Kartik) are mentioned, they are not seen as separate but obviously part of the same culture.

Sangam poetry was almost entirely lost and forgotten by the mid-nineteenth century. Luckily, a few scholars like Swaminatha Iyer painstakingly collected ancient palm-leaf manuscripts from old temples and faraway villages. In the process, it was found that there are ancient religious practices and texts which have survived from those times to this day in some regions.

All this tells us that by the late Iron Age, the people in southern India were not just aware of the rest of Indian civilization but were also fully part of it. Goods and ideas were flowing not only on the Dakshina Path but also along the coast. For some reason, Indian historians see cultural influences flowing only from the north to the rest of the country. But the fact is that these influences went both ways.

Instead of trying to split hairs over regional differences, what is amazing is that Sangam literature shows us a world that seems very familiar to us even now. For example, one of the Sangam poems gives us a glimpse of Madurai as it was under the Pandyan king Neduchelyan. We are told of the stalls near the temple selling sweetmeats, garlands of flowers and betel paan. The bazaars were full of goldsmiths, tailors, coppersmiths, flower-sellers, painters, and vendors of sandalwood. Isn't it amazing that this description would fit almost any temple-town in the south even today?

> Many think Sanskrit is a 'pure' language but in reality, it has many ideas and words from Tamil, Munda and even Greek! Many of the words that are considered to be Sanskrit which are now used in modern Tamil are actually ancient Tamil words that had been absorbed into Sanskrit! The influence worked both ways and enriched both languages.

The Old and the Gold

The world described above was at the heart of a network of merchants that extended from the Mediterranean to the South China Sea. This boom in trade happened because of an understanding of monsoon-wind patterns, a discovery that the Greeks say was made by a navigator called Hippalus.

This discovery allowed merchant ships to sail directly across the Arabian Sea rather than hug the coast. Because of this, Greek, Roman, Jewish and Arab traders flocked to Indian ports and Indian merchants made their way to the Persian Gulf, the Red Sea and even down the East African coast. We know of all these trade routes from a detailed manual called the *Periplus Maris Erythraei*, written by an unknown Greek.

According to this manual, the port of Berenike was a key hub in the trade. It was located on the Red Sea coast of Egypt and established by the Ptolemies, a Greek dynasty founded in Egypt by one of Alexander's generals.

Goods from India landed here and were taken over land to the Nile. They were then transported down the Nile in boats to Alexandria. There were other routes as well. Some ships, for example, sailed all the way up the Red Sea to Aqaba. Goods were then transported by camels and donkeys through desert towns like Petra to Mediterranean ports like Tyre and Sidon.

You must have heard of Cleopatra, the famous Egyptian queen. There is a story that when she was defeated by the Romans, she tried to escape with her family to India. She sent Caesarion, her seventeen-year-old son, whose father was Julius Caesar, to Berenike with a great deal of treasure. However, Cleopatra was captured in Alexandria and she committed suicide by snakebite. Caesarion reached Berenike and could have easily escaped to India but he was convinced by his tutors (who had probably been bribed) to return to Alexandria for negotiations. Once he came back, he was promptly murdered!

The *Periplus* tells us that ships sailing from Berenike to India went down the Red Sea to Yemen and then, dodging pirates, to the island of Socotra. This island had a mixed population of Arab, Greek and Indian traders. Even the island's name comes from Sanskrit—Dwipa Sukhadara or the Island of Bliss. This may explain why many Yemenis carry genes of Indian extract. From here, there were two major routes to India. The first made its way north to Oman and then across the Arabian Sea to Gujarat. Ships were advised to make this journey in July to take advantage of the monsoon wind.

There were many ports in Gujarat but Barygaza (modern Bharuch) seems to have been the most important. The port-town is on the estuary of the Narmada river. It's a difficult route for ships to travel because of the dangerous shoals and currents. But the local king appointed fishermen to act as pilots and tow merchant ships to the Barygaza port which was several miles upriver.

Imports into Barygaza were gold, silver, brass, copper, lead, perfumes and 'various sashes half a yard wide'. Italian and Arabian wine were imported in large quantities. The local king also 'imported' beautiful women for his harem. Exports included spikenard, ivory, onyx stone, silk and, of course, cotton textiles. As mentioned earlier, cotton textiles have always been a major export from the subcontinent.

The second route to India was a more southerly one that went across from Socotra to the Kerala coast. The most important port in this area was Muzaris (or Muchheri Pattanam) and it is mentioned frequently in Greek and Roman as well as Indian texts. A variety of goods were traded in Muzaris but the most important item of export was pepper, a spice that grows naturally in the southern tip of India. It must have been exported in very large quantities because it was commonly available as far as Roman Britain!

For a long time, historians were not sure about the exact location of this port. Excavations between 2004 and 2009 have identified it with a village called Pattanam, 30 km north of Kochi. Archaeologists have dug up a large number of Roman coins, jars and other artefacts in the area. It was a major port till it was destroyed by a big flood on the Periyar river in 1341 CE. The main trading hub then shifted to Kochi but the Muzaris area still remained important—the Portuguese and the Dutch even maintained a fort there!

> The oldest surviving structure in Muzaris is the Kizhthali Shiva temple, which is said to have been built by the Cheras in the second century BCE. The dragons carved into the steps in front of the shrine strongly remind one of the temples of South East Asia. Did this style make its way from Kerala to Java or the other way round?

During ancient times, a trade route by land from Muzaris and other Kerala ports went through the Palghat Gap (a gap in the Nilgiri mountain range near Coimbatore) to inland cities like Madurai or further on to ports in the eastern coast. Some Greek and Roman products were then re-exported to Bengal and South East Asia.

According to another ancient Greek geographer, Strabo, around 120 ships made the year-long trip to India and back in the first century CE. This probably excludes Indian merchant ships that also made this trip but in the reverse. India traded a lot with the Greeks and the

Romans at this point—this meant that there was a large flow of gold and silver coins coming in from these parts of the world. Roman writer Pliny says that India took at least fifty million sesterces (ancient Roman coins) away from Rome every year! Hoards of them have been found in excavation sites in the subcontinent, proving that this was indeed true.

At one point, the Romans were giving India so much gold that the emperor Vespasian was forced to discourage the import of Indian luxury goods and ban the export of gold to India! Over centuries of trade, India accumulated a large store of gold and silver. Even now, 25–30 per cent of all the gold ever mined is said to be owned by Indians privately though the country has very few gold mines.

Many groups of people came to India's western coast to trade or find refuge over the years. Their descendants continue to live here and in many instances, preserve ancient customs and traditions to this day. Not many know that India has one of the oldest Jewish communities in the world. It is believed that the earliest Jews came to India to trade in the time of King Solomon but after the destruction of the Second Temple by the Romans in 70 CE, many refugees settled in Kerala. St Thomas the Apostle is said to have landed in Muzaris at around this time and lived amongst this community. The descendants of the people he converted survive as the Syrian Christian community.

For fifteen centuries, the Syrian Christian community continued to observe old practices, including the use of Syriac, a dialect of the Aramaic language—the language that Jesus Christ used! Though the Portuguese tried to forcibly destroy Jewish customs, including the language, and replace them with Catholic ones in the sixteenth century, some ancient traditions continue to live on in the Syrian Christian community.

As you can see, Indian civilization is full of continuities. If Cleopatra had escaped to India, we would probably have a group that directly traced its origins to the Egyptian queen and Julius Caesar. It is still possible to experience the atmosphere of those times in the older parts of Kochi. Pepper, ginger and other spices are still warehoused and traded in the bylanes. People still sometimes use a system of hand signals, hidden from onlookers by a cloth, which evolved centuries ago. Not far is 'Jew Town' where a tiny Jewish community lives around a sixteenth-century synagogue. The Jews must have been held in high esteem by King Rama Varma for he allowed the synagogue to be built right next to his palace. Many people from this community have now moved to Israel.

What's in the Fishing Net?

Even as the western coast traded with the Middle East and the Greeks and the Romans, the eastern coast of India saw a similar increase in trade with South East Asia all the way to China. There were many ports all along the coast, including Tamralipti in Bengal, the cluster of ports around Chilka lake in Orissa, the Pallava port of Mahabalipuram and the Chola port of Nagapattinam. The importance of these ports varied over the years.

From these ports, ships sailed to Suvarnadwipa (the Island of Gold or Sumatra) and Yavadwipa (Java). Some of them sailed on further to what is now South Vietnam.

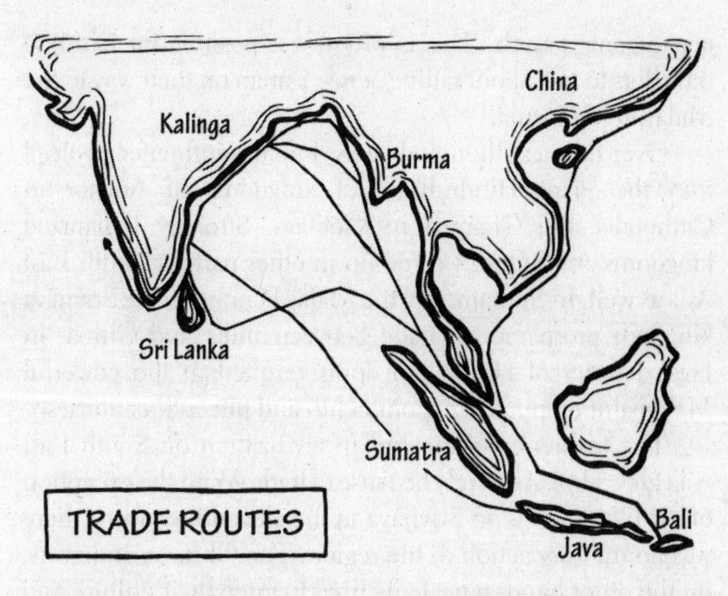

It is here, thousands of miles from the Indian mainland, that we see the rise of the first Indianized kingdom in South East Asia. Chinese texts tell us of the Hindu kingdom of Funan that flourished in the Mekong delta in the second century CE.

> According to legend, the kingdom of Funan was founded by the Indian Brahmin Kaundinya, who married a local princess of the Naga (Snake) clan. Together, they began a dynasty that ruled Funan for 150 years. The Naga or snake remains an important royal symbol in this part of the world even today.

The capital of Funan was Vyadhapura, now the Cambodian village of Banam and its main port was Oc Eo. In the early twentieth century, French colonial archaeologists found the remains of a large urban centre of houses built on stilts along a network of canals extending 200 kilometres. There were irrigation canals as well as big canals that could be used by ocean-going vessels. This is why it was possible for Chinese travellers to talk about sailing across Funan on their way to the Malayan peninsula.

Over the next thousand years, Funan's influence evolved into the great Hindu-Buddhist kingdoms of Angkor in Cambodia and Champa in Vietnam. Strongly Indianized kingdoms and cultures came up in other parts of South East Asia as well. In Sumatra and the Malay Peninsula, the Srivijaya kingdom prospered on trade between India and China. In Java, a series of Hindu kingdoms resulted in the powerful Mahapahit empire in the fourteenth and fifteenth centuries.

The influence of the Indian civilization on South East Asia grew almost entirely because of trade. With the exception of the Chola raids on Srivijaya in the eleventh century, there was no military action in the region. The Chinese emperors, on the other hand, repeatedly tried to force their culture and influence on these kingdoms through military threats. But they were not as successful as the Indians till the voyage of Admiral Zheng He in the fifteenth century.

South East Asia still bears evidence of this past. It's probably most obvious in the Hindu island of Bali but throughout the region, the influence of ancient India is alive in the names of places and people as well as the large number of words of Indian origin that are used in everyday speech.

The national languages of both Malaysia and Indonesia are called 'Bahasa' and both are full of Sanskrit words. The name itself comes from the Sanskrit word 'bhasha', meaning language. From Myanmar to Vietnam, Buddhism is the dominant religion even today. And even now, the crowning of the king of Buddhist Thailand and other royal ceremonies must be done by Hindu priests. There are more shrines to the god Brahma in Bangkok than in all of India!

India's influence is more cultural than just religious and it extends all the way to the Korean peninsula. According to the *Samguk Yusa*, Princess Huh Hwang-ok of Ayodhya sailed all the way to Korea to marry King Suro in the fourth century CE. It is said that they had ten sons who together founded Korea's earliest dynasty. The Gimhae Kim clan claims to be direct descendants of this dynasty and is still quite powerful.

It is incredible how the essence of a civilization can survive over such large distances in space and time! The Javanese perform the Ramayana in their style, against a backdrop of the ninth-century Parambanan temples. It is amazing how they evoke the landscape of a far-off time and a faraway land! The stone temples change from scene to scene. Sometimes they remind one of the rocky outcrops of Kishkindha, sometimes Ravana's palace in Lanka. A couple of hours' drive away, the sunset seen from the top of the Buddhist stupa at Borobodur still creates a magical effect even though the Buddhist chants have now been replaced by the Islamic call to prayer.

In India as well, cultural traditions continue to recall the ancient trade routes. For example, in the state of Orissa, the festival of Kartik Purnima continues to be celebrated on the day when sea merchants set sail for South East Asia. People light lamps before sunrise and set them afloat on small paper boats in rivers or in the sea. The festival is held in early November when the monsoon winds reverse. In the town of Cuttack, a large fair takes place—Bali-Yatra (meaning voyage to Bali)— around the same time. Scholars feel this marks the departure of merchant fleets for the island of Bali.

Further south, the seventh-century stone temple of Mahabalipuram still stands on the shore as if waiting for merchant ships to come home. The town, 60 km south of modern Chennai, was a busy port under the Pallava dynasty from the seventh to the ninth century CE. The existing temple complex is said to have been only one of the seven such similar complexes that once existed. It seems the others as well as numerous palaces, bazaars and grand buildings were swallowed by the sea. Local fishermen often tell tales of how their nets sometimes get tangled in such underwater structures. Historians, however, used to dismiss these stories as mere myth.

On 26 December 2004, a massive earthquake destroyed the Indonesian province of Aceh and triggered a tsunami across the Indian Ocean. About 2,30,000 people died in this tragedy. The tsunami struck India's south-eastern coast as well. However, before the waves crashed in, the sea withdrew a couple of kilometres. The residents of Mahabalipuram reported observing a number of large stone structures rising from the seabed. Then the seawaters flowed back and covered them up again. Since then, divers have confirmed that there are a number of man-made structures out in the sea though they are yet to be systematically mapped.

The tsunami also shifted the sands along the shore and this uncovered a number of other structures, including a large stone lion. Archaeologists also found the foundations of a brick temple from the Sangam period that may have been destroyed by a tsunami 2200 years ago. A second tsunami may have hit this coast in the thirteenth century. Were there six other temple complexes in Mahabalipuram? This hasn't been proved yet but once again, the memory of this culture does seem to be based on historical fact even if it hasn't been fully confirmed.

Sailing on Stitched Ships

As we have seen, the boom in overseas trade made India an economic and cultural superpower. According to Angus Maddison, a British economist, the country accounted for 33 per cent of the world GDP in the first century CE!

> Gross Domestic Product (GDP) is the broadest quantitative measure of a nation's total economic activity. It represents the monetary value of all goods and services produced within a nation's geographic borders over a specified period of time.

India's share was three times that of western Europe and much larger than that of the Roman Empire as a whole! China's share of 26 per cent of world GDP was much smaller than India's. India's population was estimated to be 75 million at that time.

What did the merchant ships in the Indian Ocean look like in those times? There were many kinds of vessels, ranging from small boats for river and coastal use to large ships with double masts for long voyages. There were also regional variations. However, they all seem to have shared a peculiar design trait: they were not held together by nails; they were stitched with rope!

Throughout the ages, travellers from outside the Indian Ocean have repeatedly commented on this odd design. This technique persisted into modern times—locally built vessels were stitched together well into the twentieth century! Apparently, there are boatbuilders who continue to do this even now. Like the Harappan ox-cart, this example shows how ancient technologies live on in India even as new ones come up.

Reconstruction of ancient Indian Ocean ship

We're not sure why the shipbuilders in the Indian Ocean region used this technique when they had access to iron nails from an early stage. Some say it may have been because of a superstition that magnetic lodestones in the sea would suck in ships which bore iron nails but this not very convincing. It's more likely that this was because these ships sailed in waters full of atolls and reefs and had to be beached in many places due to lack of sheltered harbours or due to the rough monsoon sea. This would require a hull that was flexible and did not break easily. The stitched technique provided this flexibility but it later limited the ability of Indian shipbuilding to match Chinese and then European design innovations.

So how did it feel to sail in these ships? A Chinese scholar named Fa Xian visited India in the fifth century and has left us a fascinating account of his return journey by sea. He came to India by land through Central Asia. Fa Xian spent several years in northern India studying and gathering Buddhist texts.

He then went to Tamralipti. The site of this famous ancient port, now called Tamluk, is not far from modern Kolkata. It is close to where the Rupnarayan river joins the Gangetic delta, but the old channel that served the port has silted up since. Except for a 1200-year-old temple dedicated to the goddess Kali, there is little here that suggests it was once a busy port.

In 410 CE, however, when Fa Xian visited it, Tamralipti was a port town bustling with activity. He boarded a merchant ship bound for Sri Lanka. The voyage was during the winter months when the monsoon winds would have been blowing south. The ship sailed in a south-westerly direction for just fourteen days before arriving in Sri Lanka. Fa Xian calls it the Land of the Lions—a clear reference to the mythical origins of the Sinhalese that we have talked about earlier. After all, there were never any real lions in Sri Lanka.

The Chinese scholar spent two years in Sri Lanka studying Buddhist texts before he set sail for South East Asia. He tells us that he travelled in a large vessel that could carry two hundred people. This vessel was accompanied by a smaller ship that carried extra provisions and could help in an emergency. However, after two days at sea, the ships were caught in a major storm and the larger ship developed a leak. Suddenly, there was pandemonium! Many of the merchants wanted to shift to the smaller vessel at once but its crew panicked when it saw the stampede. They cut the cables and sailed off!

This only made things worse. The merchants then threw most of their goods into the sea. Fa Xian also threw his water pitcher, washbasin and other belongings. He was afraid that the merchants would throw out his precious books but that fortunately did not happen!

Finally, after thirteen days, the storm cleared up. The crew beached the ship on a small island, possibly one of the Andaman and Nicobar Islands. The leak was found and repaired and they set sail again. But the mood remained tense because the area was full of pirates and the crew wasn't sure about its location. In the end, however, they finally set a course for Java.

The ship arrived in Java after ninety days at sea. Fa Xian, like other Chinese pilgrims who visited India to study Buddhism, saw the world in largely religious terms. The only thing he had to say about Java is that its people were Hindus and not Buddhists. This was not really accurate. After staying in Java for five months, he set sail for China on a very large merchant ship.

This ship should have been enormous because the crew alone was 200 in number! Fa Xian says he was very comfortable on the ship. It's possible that this ship had private cabins. For over a month, the ship made good progress till it, too, hit a major storm! Again, there was panic. Some of his fellow passengers apparently accused Fa Xian of bringing bad luck. They'd probably heard of his previous adventures and thought this was too much of a coincidence. Luckily for Fa Xian, a rich merchant defended him and the matter was settled.

Meanwhile, the ship's crew realized that they had been blown off-course and had no idea where they were. They had been at sea for seventy days by now and were running short on food and water. Because of this, some of the more experienced merchants decided to take control of the ship and set a new course. After sailing for twelve more days, the ship finally arrived on the Chinese coast. And thus ended one of the earliest accounts of a sea journey between India and China. This reminds us that voyages on the Indian Ocean were quite dangerous and that these ancient merchants ran enormous risks when they travelled to foreign lands.

Kings on Wheels

When Fa Xian visited India, much of the country was under the Gupta Empire, the second of India's great empires. The first of the Gupta emperors was Chandragupta I (320–335 CE) who established control over the eastern Gangetic plain with his capital in Pataliputra (now Patna). It was his son Samudragupta who dramatically expanded the empire over his forty-year rule. First, he established control over the entire Gangetic plains. Then he led a campaign deep into southern India where he defeated the kings of the region and made them submit to his rule. Having become the most powerful monarch in the subcontinent, he then performed the Vedic ritual of the Ashwamedha Yagna and proclaimed himself the Chakravartin or Universal Monarch.

Samudragupta's successor, Chandragupta II or Vikramaditya, expanded the empire westward to include Malwa and Gujarat by defeating the Sakas (or Scythians) who had ruled this area for many generations. Many of the small kingdoms and republics of north-west India also submitted to the Guptas.

There is strong evidence to suggest that the Guptas consciously modelled themselves on the Mauryans and wanted to recreate the empire that once belonged to them. Not only did two of their emperors have the same name as Chandragupta Maurya, but the Guptas also went out of their way to place their own inscriptions next to the Mauryan ones.

Much of what we know about the conquests of Samudragupta is from the inscriptions carved on an Ashokan pillar which is now in the Allahabad fort. Similarly, Skandagupta, fifth of the Gupta emperors, placed his own inscription near a Mauryan one in Girnar, Gujarat. In art and literature also, we see that the Guptas were fascinated with the Mauryans. A well-known Sanskrit play from this period, *Mudrarakshasa*, is based on the story of how Chanakya and Chandragupta defeated the Nandas and then built a mighty empire.

Many well-known scholars think that ancient Indians did not have a sense of history and that the extraordinary continuities of Indian history are all somehow accidental or unconscious. But isn't it arrogant to think that ancient people were incapable of understanding their place in history? As we can see, the Gupta monarchs clearly wanted to establish a link not just with the Mauryans but as far back as the Bronze Age. At least two of the Gupta emperors conducted the Ashwamedha Yagna or Vedic Horse Sacrifice—a ritual that was considered ancient even in the fourth century CE. The Guptas also declared themselves Chakravartins, just as the Mauryans did. And they, too, used the symbol of the wheel to convey this.

Just as the Guptas wanted to create a link with the past, they also wanted to create a link to the future. The rust-free Iron Pillar in Delhi is usually seen as an example of advanced metallurgy but it wasn't really the technology that the Guptas wanted people to marvel at. Its real purpose was to provide a permanent record of their existence. What better way to do this than to inscribe on a solid iron pillar that would never rust?

Dark and Lovely

Though the Guptas tried to emulate the Mauryans, their empire was smaller than that of the latter's. However, their two-hundred-year-rule was an economic and cultural boom. With ports on both coasts and control over major internal highways, the empire grew prosperous. We notice all this described in Fa Xian's diaries. The country must have also been well governed because though he wandered around alone for many years, the Chinese scholar does not seem to have been robbed or cheated. Later foreign travellers in India in subsequent centuries, like Xuan Zang and Ibn Batuta, all had to face armed bandits.

The Gupta emperors paid a lot of attention to intellectual and artistic excellence. It was under their rule that the astronomer-mathematician Aryabhatta worked out that the earth was spherical and that it rotated on an axis. He said the phases of the moon were due to the movement of shadows and that the planets shone through reflected light. He even worked out a remarkably accurate estimate of the circumference of the earth and of the ratio π. All this a thousand years before Copernicus and Galileo!

Emperor Kumaragupta founded Nalanda University near his capital Pataliputra. Nalanda went on to become a famous centre for Buddhist studies (although it also taught many other subjects). Further west, the Guptas established a secondary capital in Ujjain. The city became a vital trade centre in the Southern Road and also an important centre of learning for the Hindu tradition. It is said that it was here that Kalidasa, often called India's Shakespeare, composed his famous works. Ujjain is today a small town in Madhya Pradesh and it still has many ancient temples.

Fa Xian gives us an account of what it was like to visit these places in those times. He says that the cities of the Gangetic plains were very large and rich. When he visited Pataliputra, he observed the ruins of Ashoka's palace that still stood in the middle of the city after six centuries. He was so impressed by the sheer scale of the stone walls, towers and doorways that he thought they could not have been built by human hands—they must surely have been the work of supernatural creatures!

While in Pataliputra, Fa Xian observed a festival during which the people built gigantic four-wheel wagons and then built towers on them that were five storeys high. They then covered the towers in fine white linen and decorated them with canopies of embroidered silk. He says that the people placed idols of their gods within these structures and images of the Buddha on the corners of the wagons.

On the day of the festival, twenty such wagons would be pulled through the city in a grand procession. Devotees from everywhere, ranging from the royal family to the poor, were part of the festivities. They offered prayers and flowers to the gods and lit lamps in the evening. The whole city became like a fair ground with amusements and games. On this day, the rich made generous donations to the poor and physicians even held free health clinics for them.

Can you guess which festival this is? Hint—it's still celebrated in modern India! It is the Rath Yatra or Chariot Festival that is still popular among Hindus in many parts of the country. The most famous one is held in honour of Lord Jagannath in Puri, Orissa. This festival seems to have survived almost unchanged since Gupta times! The only significant difference is that earlier Buddhists used to actively be part of this festival. It looks like Buddhism and Hinduism were seen to be religions that

went together despite the disputes among scholars about their respective teachings. This relationship is still alive among the Dharmic religions. The Nepali Hindus, for example, pray at Buddhist shrines just as Buddhist Thais commonly pray to the Hindu god Brahma and Punjabi Hindus visit Sikh gurudwaras.

The *Kamasutra* (the Treatise on Love) was written during the rule of the Guptas. It gives us a vivid picture of what life was like for the rich and the idle in those times. A lot of importance was given to personal grooming—a man is required to have an oil massage, bathe, shave, apply perfumes and clean the sweat from his armpits. After lunch, a man ought to entertain himself by teaching his parrot to talk or by attending a cock-fight or ram-fight. Then, after a nap, he should dress up and head for the salon. Evenings were to be spent with friends, in the company of courtesans. The *Kamasutra* also describes picnics in the same way, with great detail on how people ought to enjoy themselves and have a good time! The lives of the idle nawabs of nineteenth-century Lucknow and the Bengali zamindars of early twentieth-century Kolkata were quite similar. The Page 3 socialites in cities like Mumbai and Delhi still enjoy such a lifestyle!

What did people and their world look like in this period? We can find out by studying the carvings and paintings in the Ajanta and Ellora caves in Maharashtra. These caves were constructed during the rule of the Vakatakas, close allies of the Guptas. The paintings may not be of ordinary people but they still reveal glimpses of courtly life and what Indians of this period idealized. One of the most striking things is that most of the people in the paintings are very dark-skinned. It looks like ancient Indians had a preference for dark skin. There is a lot of other evidence also to support this. For example, Krishna is considered to be the most handsome male in Hindu tradition and his name literally means 'the dark one'. He became blue-skinned because of medieval artists who depicted him that way. Similarly, Draupadi, the wife of the Pandavas in the Mahabharata, is also described as being very dark.

Even in the medieval period, dark skin seems to have been the preference among Indians. Marco Polo says in his comments about India that the darkest man was the most highly esteemed and considered better than others who were not so dark. The gods and their idols were black while the devils were shown as white as snow. You only need to look at the idol of Lord Jagannath in Puri to see what Polo meant!

It is not clear when things changed and fair skin came to be preferred in India, but we should remember that the traditional Indian aesthetic was very different from how we see it now. Not a very good thing for those who want to sell skin-whitening creams to modern Indian consumers! Of course, it's not only Indians who have changed their tastes over the years. Just a few generations ago, Europeans thought pale white skin was so attractive that women were willing to risk poisoning by using an arsenic-based compound to whiten their skin. Today's Europeans risk skin cancer from too much sunbathing! Whichever way you go, you basically can't win.

Holy Water

By the first half of the sixth century, the Gupta Empire gradually began to crumble. There were internal problems and repeated attacks by the Hunas (White Huns) from the North West. Taxila, the famous centre of learning, where Chanakya had once taught, was attacked by the Hunas around 470 CE. Over the next few decades, they pushed the Gupta defences back into the Gangetic valley. Some part of the empire remained for several generations but their days of glory were gone.

After this, a number of powerful kingdoms rose and fell in northern India—the Palas of Bengal, Harsha of Thaneshwar and so on. The city of Pataliputra slowly went into decline and was replaced by Kannauj (now a small town in Uttar Pradesh). Both Nalanda and Ujjain, however, remained important centres of learning. Thinkers like Bhaskara and Varahamihira made great contributions to mathematics and astronomy. Chinese scholars continued to visit India to study Buddhism. Xuan Zang was one such scholar who visited India in the seventh century, over two hundred years after Fa Xian.

Like Fa Xian, Xuan Zang also made his way to India through Central Asia. He spent over a decade in the subcontinent during which he criss-crossed the Gangetic plains and went as far east as Assam. He even spent two years studying at Nalanda, which was then at the height of its fame.

One of the places Xuan Zang visited on his travels was Allahabad (then called Prayag). This town is near the Triveni Sangam or the 'Mingling of Three Rivers', considered a very sacred site in Hinduism. Two of the rivers are obvious—the Ganga and the Yamuna. The third is supposed to be the Saraswati which is said to flow underground and join the other two at this place. Though this ancient river has vanished, people still remember it.

For a Hindu, the sins of a lifetime are believed to be washed away by taking a dip in this confluence of rivers. It is especially auspicious to do this at the time of the Kumbh Mela which is held here every twelve years. This is part of a four-year cycle by which this event is held in turn in Ujjain, Haridwar, Nasik and Allahabad. However, it is the great Kumbh Mela of Allahabad that is the largest and most prestigious. The last time this festival was held in Allahabad in 2013, close to a hundred million people participated in it!

Xuan Zang tells us that large numbers of people participated in the festival in the seventh century, including the rulers of different countries and even Buddhists. He also describes the rituals of sadhus or ascetics. A large wooden column was erected in the middle of the river and the sadhus would climb it. At sunset, they would hang from the column with one leg and one arm while stretching out the other leg and arm into the air. They would then stare at the setting sun. The wooden pole and the particular ritual are no longer around but there are still ash-covered sadhus who visit the Kumbh Mela.

Though these continuities are still present, it looks like the economic and cultural centre slowly shifted to the south. Even militarily, the southern kings had become more powerful. When Xuan Zang visited India, the northern plains had come under the rule of Emperor Harsha but he was soundly defeated by the Chalukya king Pulaksen II when he tried to extend his empire into the Deccan.

The southern kingdoms had become so powerful because of trade—with the Indianized kingdoms of South East Asia as well as with the Persians and the Arabs who had replaced the Romans in the west. The kingdoms of the south were aware of the importance of trade and actively encouraged it. They were not afraid to fight to keep the trade routes open. The most famous examples of this are the Chola naval expeditions to South East Asia in the eleventh century.

The Cholas were an ancient dynasty and are even mentioned in the Ashokan inscriptions. In the ninth to the eleventh century CE, they created an empire that covered most of peninsular India and briefly extended to the banks of the Ganga. The empire even included Sri Lanka and the Maldives! They had very good relations with kingdoms of South East Asia. Inscriptions on both sides show that there were large merchant communities and that the kings exchanged emissaries and gifts often.

But a problem probably arose because the Cholas began to create direct trade links with the Song Empire in China. Records show that the Cholas and the Chinese exchanged a number of trade delegations in the early eleventh century. Even before this, there was trade between India and China. A large Indian merchant community had been established in Guangzhou and there were even three Hindu temples functioning there. But there appears to have been a big boom in trade once the Song Empire and the Cholas had established direct links.

The Srivijaya kings who functioned as middlemen did not like this. They started to tighten controls and imposed heavy taxes on ships passing through the Straits of Malacca. An Arab text tells us that the Srivijaya kings demanded a levy of 20,000 dinars to allow a Jewish-owned ship to continue to China! This was a serious matter and the Cholas were not amused. They conducted a naval raid against Srivijaya in 1017 CE and then a more substantial expedition in 1025. This was a rare example of Indian military aggression outside the subcontinent. It did not last long however; a few decades later, the Cholas and the Srivijayas sent joint embassies to the Chinese.

Just as India exerted its influence on South East Asia, there were many influences from the South East that came to the subcontinent. For example, the University of Nalanda grew the way it did because of the strong financial support it received from the Srivijaya kings. South East Asian kingdoms like Angkor, Majapahit and Champa accepted Indian influences and built on them, innovating and adapting them to their own culture. Moreover, the Indonesians independently conducted their own maritime expeditions. From the third to the sixth century

CE, they crossed the Indian Ocean and settled in Madagascar in large numbers. The first inhabitants of Madagascar thus came from distant Indonesia and not nearby Africa! The descendants of those Indonesian settlers still form a significant part of the population in Madagascar and the Malay language contains strong influences of dialects from Borneo.

A Chain of History

Many scholars say that ancient Indians wrote only one formal history—Kalhana's *Rajatarangini* or River of Kings, a history of the kings of Kashmir written in the twelfth century. They say that this shows how Indians did not have a sense of their history or of the continuity of their civilization. But the lengthy records maintained by different traditions like the Vanshavali of Nepal or the Burunjis of Assam that keep track of family lineages document the great importance given to remembering the past.

Kashmir's Kalhana saw himself as a link in this chain. He tells us that he read the works of eleven earlier historians and inspected numerous temple inscriptions and land records. He even criticized fellow historian Suvrata for leaving out details and making his history too short! Most of the works Kalhana studied may have been lost but they all clearly existed. Kalhana's history is followed by three other works that continue the chronicle down to the time of the Mughal emperor Akbar.

When Akbar conquered Kashmir in the sixteenth century, he was given a copy of *Rajatarangini* that was translated to Persian. A summary was then included in the *Ain-i-Akbari*, the chronicles of Akbar's own rule, in order to link him into this historical chain.

Kalhana's history is not just about kings and battles; it also contains an interesting account of how human activities changed the landscape of Kashmir. He tells of the minister Suyya who carried out many major engineering works during the reign of Avantivarman in the ninth century. The landslides and soil erosion had led to a great deal of rubble and stone being deposited in the Jhelum river and this was disturbing its flow (the Kashmir floods of 2014 demonstrated the risks). This rubble was removed and embankments were built. The landscape was restructured to human use as dams created new lakes while old ones were drained to clear the way for cultivation. It is suggested that Suyya may have significantly altered the course of the Jhelum and Indus rivers. It looks like much of the 'natural' beauty of Kashmir may actually be due to thousands of years of human intervention!

5
Sinbad the Sailor

We're now in the part of the story when Emperor Harsha was building his empire and Xuan Zang, the second Chinese scholar we discussed, was setting off on his long pilgrimage. Around this time, there was a former merchant called Muhammad whose actions would have a great impact on the world over the years.

By the time Prophet Muhammad, the founder of the Islamic religion, died in 632 CE, he already controlled much of the Arabian Peninsula. Within a century, his followers created an empire that stretched from the Iberian peninsula to Central Asia. In the eighth century, the Arabs gained some control in Sindh by defeating Raja Dahir.

This victory in Sindh, however, did not seem to have had much effect on the Indian heartland. The Arabs tried to expand further but they were warded off by the Rashtrakuta and the Gurjara-Pratihara kingdoms (the latter gave their name to the state of Gujarat). Arabian records from these times talk of the excellence of the Indian cavalry. The emerging Rajput military class actually seems to have made

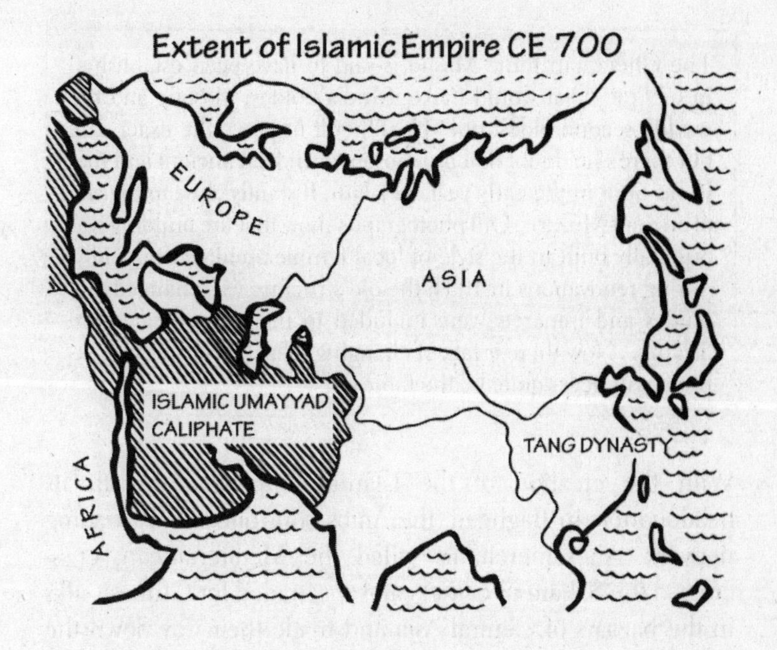

counter-attacks of its own on the Arab Peninsula and much of Afghanistan continued to be ruled by the Hindu Shahis well into the tenth century. For the first several centuries of Islam, India's interaction with the religion was not about conquests but about trade.

A Sword Like Water

The Arabs had been actively trading with India even before the origin of Islam. In the early seventh century, the ports along the western coast were regularly visited by Byzantines, Persians, Yemenis, Omanis and even Ethiopians. There were merchants from the Mecca region too. Muhammad probably knew many of the merchants who visited India.

The Cheraman Juma Masjid is said to have been established in 629 CE. That would make it India's oldest mosque and the world's second oldest one! It's difficult to prove the exact date but there's no doubt that the mosque is indeed ancient and that it was built in the early years of Islam. It stands close to the site of ancient Muzaris. Old photographs show that the building was originally built in the style of local temple architecture. Sadly, during renovations in 1984, the old structure was changed and domes and minarets were included to make it appear more 'Islamic'. Now there is talk of changing it back to attract tourists but it will never quite be the same.

With the creation of the Islamic empire, and with its headquarters in Baghdad, the Arabs controlled a vast trading network. Arab merchants sailed the Mediterranean, criss-crossed the Sahara in camel caravans, traded for Chinese silks in the bazaars of Central Asia and made their way down the East African coast in search of slaves. Do you remember the remarkable adventure tales of Sinbad the Sailor? Even if the tales of *One Thousand and One Nights,* which belong to this age, are fictional, they quite clearly convey the spirit of the era!

The Iraqi port of Basra became the most important trading centre of the empire because of its nearness to the capital. Indian goods and merchants dominated this market so much that the Arabs spoke of Basra as 'belonging to al-Hind'. The commodities of trade included perfumes, spices, ginger, textiles and medicinal substances. After the Arabs conquered Sindh, large numbers of slaves were also brought in from there.

Interestingly, the most important Indian export of the period was the steel sword. Indians were famous at that time for the quality of their metal goods and the swords used by early Muslim armies were often of Indian origin. This remained true even at the time of the Christian Crusades.

The famous 'Damascus Sword' was either imported from India or was made using Indian techniques.

> A Damascus steel sword is very distinctive to look at. It has patterns in the steel (because of banding and mottling) that give it the appearance of flowing water. Such blades were famous for being tough and resistant to shattering. They could be honed to a very sharp, durable edge.

Just as in South East Asia, there were large numbers of Indian merchants who lived along the ports of the Arabian Sea and the Persian Gulf as well as in inland trading towns. Similarly, Arab merchants also came in large numbers to the ports along India's western coast. The famous Arab historian and geographer Masudi tells us that Indian kings welcomed the traders and allowed them to build their own mosques. He tells us of a particularly large settlement of ten thousand Muslims in the district of Saymur where immigrants from Oman, Basra, Siraf and Baghdad had permanently settled.

Farther south, there were a number of Arab settlements in Kerala where the Arabs mixed with local converts. Their descendants, the Moplahs or Mappilas, form a quarter of the state's population today. Since the 1970s, however, these people (and others from Kerala), have been going in large numbers to work in the oil-rich Arab states because of changing times.

Meanwhile, farther north, Gujarat became home to the Parsis, followers of the Zoroastrian tradition. As discussed in Chapter 2, the origins of the Zoroastrian tradition are closely linked to the Rig Vedic people. For fifteen hundred years or more, Zoroastrianism was the dominant religion in Persia. But its influence reduced greatly after the Islamic conquest of the region. Religious persecution (the punishment of people who did not follow the faith of the conquerors) was quite common. A small group of Zoroastrians therefore escaped to Gujarat in the eighth century. According to the *Qissa-i-Sanjan* (an epic poem that the Zoroastrians believe is an account of their early years in the Indian subcontinent) the local Hindu king allowed them to settle on his land on the condition that they give up their arms and adopt the local language—Gujarati. They were, however, given full freedom to follow their religion.

The descendants of these people migrated to Mumbai in the nineteenth century to repair and build ships for the British. Some of them sailed to British-controlled Hong Kong and made large profits from participating in the opium trade with China. The Parsis brought back this money and built large commercial and industrial enterprises in Mumbai. They are still one of India's most successful business communities.

Apart from slaves and merchants, there were several other Indian groups in the Middle East during this period, including mercenaries. A mercenary is a person who fights for money and can be hired by anyone to be on their side during a war. He has no particular loyalty to anyone or any group. According to the oral tradition of the Mohyal Brahmins of Punjab, some of their ancestors died fighting for Hussein in the Battle of Karbala, Iraq in 680 CE. This is why this particular group of Hindus, also known as Husseini Brahmins, still join Shia Muslims during the ritual mourning of Muharram every year.

At about the same time, another group from central India travelled west, across the Middle East, to Europe. We know them today as the Gypsies or the Roma. Language and culture have always indicated that there is a link between the Roma and India, and genetic studies have now confirmed it. We don't know why this group left the subcontinent; it is possible that they were a group of stranded soldiers from Gurjara-Pratihara armies fighting the Turks and Arabs in

Sindh. They were probably always a nomadic group and some circumstances may have made them move westwards.

> In 1971, at the World Romani Conference near London, the Roma adopted a blue and green flag for their nomadic nation. At the centre of the flag, they placed a wheel—the symbol of the Chakravartin. We must admit that the Roma probably have the greatest claim to this symbol now. Their wheels can truly roll in any direction!

The exchange of goods, people and ideas was happening within the subcontinent just as it was happening with the outside world. For example, the philosopher Adi Shankaracharya, who was from Kerala in the extreme south, travelled all over the country in the eighth century. His ideas became influential within the subcontinent and beyond it. Similarly, the Shakti tradition associated with the worship of Goddess Durga and her incarnations started in the eastern regions of Bengal and Assam. However, by the medieval period, there were fifty-two shakti-peeths or pilgrimage sites related to this tradition, which were spread across the subcontinent—from Kamakhya in Assam to Hinglaj in Baluchistan, and from Jwalamukhi in Himachal Pradesh to Jaffna in Sri Lanka. There are even Shakti temples in South East Asia. The ninth-century Prambanan temple complex in central Java, Indonesia, has a shrine dedicated to the goddess Durga slaying the demon Mahishasura. This exquisitely carved idol would not look out of place in the annual Durga puja festival in modern Kolkata!

Watch Out, It's the Turks!

In many ways, life in the subcontinent till the beginning of the eleventh century remained more or less the same as earlier times. Maritime trade continued to flourish in the southern ports, and foreign scholars came in large numbers to study in Nalanda. There had been changes in architecture, technology and style, but the cities of the subcontinent would have been quite familiar to a visitor from a thousand years earlier. However, all this was about to change.

In the late tenth century, the Turks began to invade the Hindu-Buddhist kingdoms of Afghanistan. In 963 CE, they captured the strategically important town of Ghazni. From there, they conquered the Hindu Shahi kingdom of Kabul and pushed them back into Punjab. The Shahis fought back for decades but on 27 November 1001, they were defeated by Mahmud of Ghazni in a battle near Peshawar. The Shahi king Jayapala was so distraught that he gave up his throne to his son and stepped on to his own funeral pyre. The Shahis continued to fight the Turks but their power had reduced considerably.

Over the next twenty-five years, Mahmud made seventeen raids into India, many of them directed at wealthy temple towns such as Mathura and Nagarkot. His most infamous raid was against the temple of Somnath, Gujarat, in 1026 CE. It is said that this attack left over fifty thousand people dead and about twenty million dirhams worth of gold, silver and gems were taken away. The Somnath temple has since been destroyed and rebuilt many times but it is Ghazni's raid that is still remembered most vividly.

The Somnath temple as it stands today was built in the early 1950s. Its reconstruction was one of the first major projects started shortly after India became a republic. Standing right on the seashore, the temple is a wonderful place to watch the sun set. But something of Ghazni's massacre still seems to linger in the air. Barely half a kilometre away is the spot where the Pandava warrior Arjuna is said to have conducted the last rites of Krishna. Three rivers meet in the sea here—one of them is named Saraswati.

The Turks were eager to gain wealth and spread their religion but there was also another important interest they had in mind—the capture of slaves. Over the next few centuries, hundreds of thousands of Indian slaves—particularly from West Punjab and Sind—were marched into Afghanistan and were then sold in the bazaars of Central Asia and the Middle East. They were unused to the extreme cold of the Afghan mountains and died in such large numbers that the range came to be known as the Hindukush or the 'Killer of Hindus'.

These raids and invasions were not met with a strong response as there would have been in the times of the Mauryans or the Guptas. The last great Hindu empire of North India—that of the Gurjara-Pratiharas—was reduced in power and the heart of the civilization had shifted south to the Vindhyas. The most powerful Indian kingdom of that time was that of the Cholas, who ruled in the far south and were not much concerned with what was happening in the North West.

Meanwhile, freed from the political and cultural domination of the Gangetic plains, central India experienced a cultural and economic boom. This was the age of the remarkable Raja Bhoj, the warrior-scholar who ruled much of central India and of the Chandelas of Bundelkhand, who built the temples of Khajuraho.

Raja Bhoj is not given much importance by historians but central India is full of stories and ballads about him. How much of these are true? It's hard to say but one cannot deny his importance to this region. Raja Bhoj rebuilt the Somnath temple, fought against many Turkish raids and built one of the largest forts in the world at Mandu in Madhya Pradesh. But the most visible of his achievements is the huge lake that he created using an earthen dam in Bhopal, a city that is named after him. Before the Bhopal gas tragedy of 1984, the city was best known for this body of water. It shows the skills of the medieval engineers.

The lake still stands, after all these years, proving that big dams work well in the Malwa plateau unlike the Himalayas where tectonics and silt can make them risky to build.

Farther north, the Chandelas of Bundelkhand, who once used to submit to the Gurjara-Pratiharas, carved out a small but powerful kingdom for themselves after the latter lost their powers in the tenth century. They celebrated their successes by building the famous temple of Khajuraho, which is now a UNESCO World Heritage Site. The Kandariya Mahadev temple, the largest in the complex, is said to have been built after the Chandelas fended off Mahmud Ghazni himself!

There are many striking sculptures in Khajuraho that depict lions or lion-like dragons. Many of them are shown locked in combat with a Chandela king or warrior, including female warriors. Just like the Mauryans, the Chandelas also liked to use the lion as a symbol of power. There are no tigers anywhere among the sculptures, but the Panna Tiger Reserve is just a twenty-minute drive away from Khajuraho. Was this a region dominated by lions in those times? Or did the Chandelas simply think that tigers were not symbolic of royal power?

For a century and half after Mahmud's raids, the Turks were mostly restricted to West Punjab. But in 1192, Muhammad Ghori defeated Prithviraj Chauhan, the Rajput king of Delhi and Ajmer, in the Second Battle of Tarain (150 km from Delhi in the modern state of Haryana).

The Turks occupied Delhi and then invaded the rest of India. By 1194, Varanasi and Kannauj were captured and ransacked. Kannauj never really recovered from this attack. Within a few years, the University of Nalanda was destroyed by Bakhtiyar Khilji. Its library was torched and most of its scholars were put to death.

While there were many Brahmin scholars at Nalanda, most of them who were killed were probably Buddhist monks. Some of those who survived stayed on but most fled to Tibet where they continued following their traditions till the Chinese takeover of the mid-twentieth century. After Nalanda, Bakhtiyar Khilji attacked and completely destroyed Vikramshila, another famous Buddhist university. Buddhism was already losing its influence in the subcontinent and these attacks led to its collapse. In 1235, Sultan Iltutmish attacked Ujjain, once the secondary capital of the Guptas and a major centre for the study of mathematics, literature, astronomy and Hindu philosophy. It was around the same time that these universities were being destroyed that the University of Oxford was being established on the other side of the planet.

By the end of the thirteenth century, armies led by generals like Malik Kafur made raids into the deep south. This was a bloody period in Indian history—ancient cities, universities and temples were ravaged and millions were probably killed. And so ended the second cycle of urbanization that had begun in the Gangetic plains during the Iron Age.

Some sparks from the old times remained alive in the city of Vijayanagar in the far South and in the even more faraway kingdoms of South East Asia. However, India was starting on a new cycle of urbanization which was influenced a great deal by Central Asia and Persia. This book can't go into all of it in detail, so we will look at the next cycle of urbanization largely through the evolution of its most prominent city—Delhi.

Many, Many Delhis

People have been living in and around Delhi from the Stone Age—tools and other objects made of stone have been found in the ridges of the Aravalli range. The Rig Vedic people would have been familiar with this region since it is in the eastern corner of the Sapta-Sindhu. Remains from the late Harappan age have also been found here. It could be that this was one of the places where the Harappans settled when the Saraswati dried up.

Since this time, cities have been built, abandoned, destroyed, and rebuilt many times. Some say Delhi has been built eight times. Others say the number is sixteen. In the last 150 years alone we have seen a full cycle. After Delhi was sacked by the British in 1858, its urban population fell to a mere 1,54,417 people. Today, the National Capital Region is home to twenty million people and it is growing rapidly.

The terms 'Old Delhi' and 'New Delhi' have meant different things at different points in time. When William Sleeman visited the city in 1836, he called Shahjehanabad what we now call 'Old Delhi','New Delhi'. To him, Old Delhi was the ruins that were scattered from Mehrauli to Purana Qila—the area that we now call New Delhi!

As we discussed in Chapter 3, many of the events of the Mahabharata are said to have happened in the area around Delhi. Indraprastha, the capital of the Pandavas, was situated in Delhi along the banks of the Yamuna. Archaeological excavations inside the Purana Qila fort, under which this site was believed to have been, bear evidence of a Late Iron Age settlement. The settlement seems to have been occupied till the Gupta period. There is a small museum in Purana Qila which displays photographs and artefacts from the excavations.

However, it is impossible to prove that this was indeed the city mentioned in the epic. Nothing found so far obviously matches the descriptions of Indraprastha. There is no palace or audience hall that would have made the Kauravas jealous. It could be that archaeologists have not yet found it or that they were swept away by a flood on the Yamuna. But what we can say is that Delhi was an important settlement from very ancient times. An Ashokan rock inscription discovered near the Kalkaji temple in 1996 suggests that Delhi may have included several habitations other than just the one at Purana Qila.

The first Delhi of which we know was built by the Tomar Rajputs. They made it their headquarters in the eighth century. Their first settlement was at Suraj Kund in the extreme south of Delhi. Water supply was a big concern then (as it is now) and a large stone dam that was built by the Tomars still stands. The nearby village of Anangpur recalls the name of Raja Anang Pal, a Tomar king. A stream from the dam feeds a stepped tank that was probably linked to a temple of the sun god—hence the name Suraj Kund or Pool of the Sun. The lake is often dry now because of urbanization and illegal quarrying.

In the eleventh century, the Tomars moved farther west and constructed a large fort—Lal Kot or the Red Fort. Shah Jehan's seventeenth-century Red Fort was not the first to have that name. To mark his place in history, Anang Pal also added his name to the Iron Pillar.

A century later, the Chauhans of Ajmer took control of the city and further expanded it. It now came to be called Qila Rai Pithora and was the capital of Prithviraj Chauhan. Large sections of the walls of this city can still be seen near Mehrauli village though almost nobody visits them. From the top of the walls, you can see towers and other structures including a major gateway. The urban landscape of Delhi can be seen beyond the trees even as the Qutub Minar looks on sternly.

When the Turks captured Rai Pithora, they made it their Indian headquarters and began to remodel it for their own use. The Qutub Minar complex, a UNESCO World Heritage Site, has some of the oldest Islamic buildings in northern

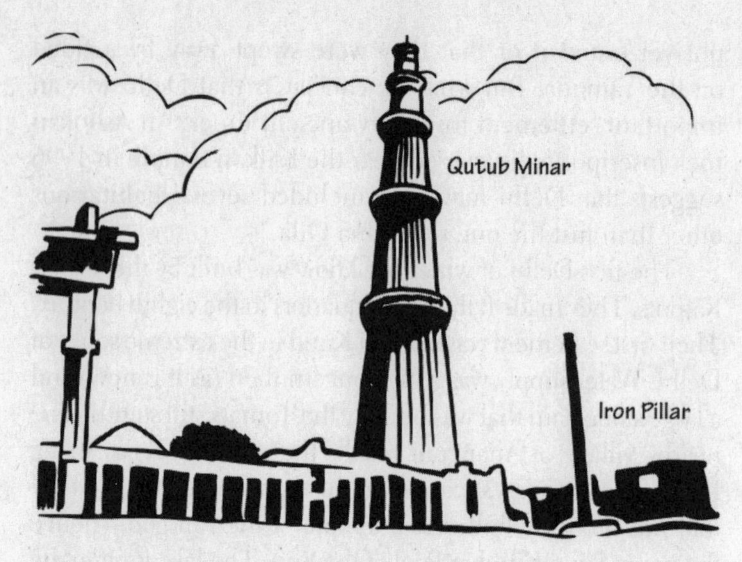

India. At its centre is the mosque built by Ghori's slave-general, Qutubuddin Aibak. An inscription to the east gate of the mosque says that it was built on the remains of twenty-seven demolished Hindu-Jain temples. Destroyed idols can still be seen among the columns. The Iron Pillar, however, continued to stand on one side of the mosque courtyard. Was it that the new rulers wanted to use this ancient symbol of power for themselves? It is also possible that Qutubuddin wanted to let it stand in the shadow of his own great pillar—the Qutub Minar, a 72.5m-high stone.

The Qutub Minar is a truly impressive structure even by modern standards. When the Moroccan traveller Ibn Batuta visited Delhi over a century later, he was stunned by the height of the tower as well the unique metallurgy of the pillar. He tells us that a later sultan wanted to build a tower twice as high but the project was abandoned. The remains of that sultan's attempt still stand. The sultan was Alauddin Khilji.

At the beginning of the fourteenth century, Alauddin Khilji built the new fort of Siri at the site of a military camp north-east of the existing city. The main urban centre continued to be in the old city but the sultan, who was worried that he would be assassinated, felt safer within the new fort. But a few years after it had been built, Siri was raided by the Mongols. Alauddin managed to push the Mongols back and then took care to make the fort stronger. This was a wise move because the Mongols were soon back! They succeeded in capturing the main city and looted it. However, Alauddin was safe in Siri for months till the Mongols decided to return.

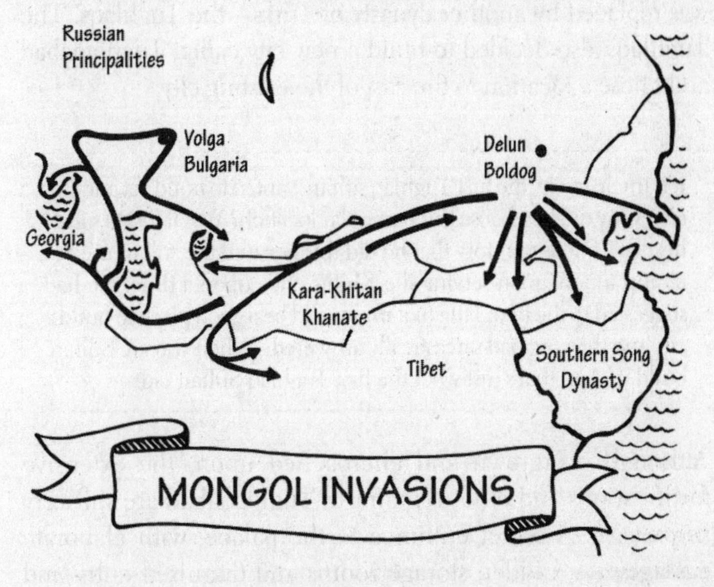

There are only a few stretches of walls and other buildings that remain of Siri now. Its site is now occupied by the urban village of Shahpur Jat. This is one of the many villages that live on in modern Delhi and is now dotted with many small offices and designer workshops. The rest of Siri is covered by the Asian Games Village that was built to house athletes for the event in 1982.

As we've seen, water supply has always been a major problem in Delhi. In Khilji's times, he built a large reservoir called Hauz Khas which still exists and is surrounded by a beautiful park. Overlooking the reservoir are the remains of an old Madrassa (Islamic religious school) built by a later sultan, and the urban village of Hauz Khas. Since the 1990s, this village has changed a lot. Now, there are a number of expensive boutiques and trendy bars there—pretty ironic because Alauddin was a severe man who strictly controlled the prices of all goods and did not permit the buying or sale of alcohol!

The Khilji dynasty did not survive for long after Alauddin. It was replaced by another dynasty of Turks—the Tughlaqs. The Tughlaqs also decided to build a new city called Tughlaqabad and chose a location to the east of the existing city.

It is unclear why the first Tughlaq sultan wanted to build yet another city. Why did he choose this particular location? Was it just to satisfy his ego? There is a story that he did this because the sultan used to be just a nobleman serving the Khiljis once upon a time. He had suggested to the then king that this would be a good place to build a city and the king had sarcastically answered, 'When you are Sultan, build it.' And that's just what the first Tughlaq Sultan did!

Although overgrown and encroached upon, the extensive fortifications and other remains of Tughlaqabad are still very impressive. A secret entrance to the palace, with elaborate passageways, hidden storage rooms and disguised entry and exit, were discovered in the 1990s. The exit is a small opening on the outer wall which looks like a drain. However, though the city looks impressive, it was occupied only for a few years. Probably because of water supply problems!

Muhammad Tughlaq was the second sultan of this dynasty. He decided to shift his capital a thousand kilometres south to

Daulatabad in 1326 CE. The fort was located in a strategically crucial position on the Southern Road or Dakshina Path. It would be easy to make raids from here into southern India. The sultan's decision made sense but he also insisted that every single person living in the old capital move with him. There is a terrible tale of how an old beggar who couldn't make the journey was tied to a cart and dragged along for forty days. Obviously, he did not reach the new capital alive.

After putting the people through such trouble, the sultan changed his mind! The entire population was then made to walk back to Delhi, the old capital. Muhammad decided to expand Delhi and invited settlers from the rest of his empire and from Central Asia. He built a set of walls that connected the old city of Lal Kot with the Khilji fortress of Siri. This was a very large area and it would become the next city of Delhi—Jahanpanah. The older cities continued to have people living in them even as capitals shifted. Even some parts of the abandoned Tughlaqabad were used for storage and to house soldiers. Jahanpanah probably contained open areas and even farming communities within the walls. A massive new palace-complex was built. It is all of this that Ibn Batuta saw when he visited Delhi.

Ibn Batuta is one of the greatest travel writers of all time. Originally from Tangier in Morocco, he travelled across the known world through North Africa, the Middle East, Central Asia and India, and eventually to China. He then made his way back to Morocco before orally recording the story of an adventure that had lasted for almost thirty years. When he arrived in Delhi, Muhammad Tughlaq ruled from his new palace in Jahanpanah. This was the time when the sultan was trying to build up the prestige of his court by bringing in learned Muslim scholars. Ibn Batuta thus became part of the court.

Batuta has left us a very vivid description of the Tughlaq court. One had to pass through three gates to enter the sultan's audience hall. There was a platform in front of the first gate where the executioners sat. When a man was sentenced to death, he would be executed outside this gate and his body would be left there for three days as a warning to others. At the first gate sat a number of trumpeters and pipe-players. Whenever an important person came, they would blow the trumpets and loudly announce the person's name. Between the first and second gates, there were platforms occupied by large numbers of palace guards.

At the second gate sat the royal ushers in gilded caps, the chief usher wielding a golden mace. Inside this gate was a large reception hall where people could sit as they waited for their turn. The visitor would then walk up to the third and final gate where scribes entered the person's name, time of arrival and other details. A nobleman who didn't attend the court for more than three days without a valid reason was not allowed beyond this point without the sultan's permission. Beyond this gate was the main audience hall, a large wooden structure of 'a thousand pillars'.

The sultan sat on a raised throne supported by cushions. An attendant stood behind the king and flicked away the flies that may dare to disturb him! And there probably were many flies—you cannot leave dead bodies lying around the front porch for days and not expect pests! In front of the sultan were the members of the royal family, the nobility, the religious leaders, the judiciary and so on. Each person was given a position according to their status. When the sultan arrived and sat down, the whole court rose and shouted 'Bismillah!' A hundred of his armed personal guards stood on either side of the throne. Clearly, the sultan was taking no chances.

Most of the nobility and senior officials were foreigners—Turks but also Khurasanis, Egyptians, Syrians and so on. Ibn Batuta says that Muhammad Tughlaq always gave the high positions to foreigners. This was not unusual. The Turks were

Muhammad-Bin-Tughlaq

an army of occupiers and the Indians—both Hindus and local Muslims—would have been treated with contempt. Indians were probably allowed into the walls of medieval Delhi only as slaves and menial workers. This attitude only changed under the Afghan Suris in the mid-1500s and became a common practice under Mughal emperor Akbar. It was a slow change. Even under Akbar, most of the nobility was foreign-born.

From time to time, the Tughlaq sultan held banquets. All the guests were seated by rank. Each person was first given a cup of 'candy-water' and they had to drink this before they began the meal. Then the food was brought from the kitchen in a procession headed by the Chief Usher carrying a golden mace and his deputy holding a silver mace. As they walked by, they would cry out 'Bismillah!'

And what was the food like? There were rounds of unleavened bread, roast meats, chicken, rice, sweets and . . . samosas! After the meal, each person was given a tin cup of barley water for the stomach. After this, everyone got paan (betel leaves and areca nuts). When it was all done, the chamberlains cried 'Bismillah!' and everyone stood up. We are not sure where exactly Muhammad Tughlaq's palace stood but the most likely place is the ruins of Bijay Mandal, quite close to IIT Gate. Very few tourists today visit the site. Just behind the Bijay Mandal complex is the Begumpur Mosque which may have once been the imperial mosque.

Ibn Batuta became afraid of the sultan over time. He eventually attached himself to an embassy to China and fled the country. But as the embassy made its way south, it was repeatedly attacked by bandits. Ibn Batuta was captured and almost killed but was finally set free. The fact that even an imperial embassy was not spared by bandits on a major highway tells us how chaotic things had become under Turkic rule.

After many more adventures, Ibn Batuta reached China. But though he is still famous in the Arab world, apart from historians and scholars, his adventures are barely remembered in India. His memory is now limited to a mention in a somewhat silly but popular Hindi song about his shoe!

The Tughlaq dynasty had built three capitals by now—Tughlaqabad, Daulatabad and Jahanpanah. But they were still not satisfied! Feroze Shah Tughlaq, who came after Muhammad, was an even more enthusiastic builder. He constructed many new structures as well as repaired many of the old ones. He also extended the city northwards by building a fortified palace-complex along the Yamuna called Feroze Shah Kotla.

The ruins of Feroze Shah Kotla are now near the busy ITO crossing, just behind the offices of India's leading newspapers. The architects of later Delhi often took their building material from this structure but the site still contains a three-storeyed pavilion topped by the Ashokan pillar that had been carefully brought here by the sultan. The complex is said to be inhabited by djinns (spirits), and people, mostly Sufi Muslims, often come here and light lamps to console them and also ask for favours. You can see small offerings and walls blackened by the smoke from the lamps. Some of the believers tie colourful strings to the grill put up by the Archaeological Survey to protect the Ashokan pillar—modern Indians thus pay their respect to an ancient imperial pillar in order to communicate with medieval spirits!

Feroze Shah came to the throne when he was almost a middle-aged man and he ruled till his death. He died after several years of illness at the age of eighty-one. As with Ashoka, his empire was already weakening towards the end and he was followed by rulers who could not manage the empire effectively. The Turko-Mongol raider Taimur the Lame (also known as Tamerlane) took this opportunity to sweep into the country from Central Asia in 1398. He defeated the Sultan's army easily and entered Delhi. He spared Muslim territories but everything else was looted or destroyed. The entire Hindu population was either killed or taken away as slaves. Taimur later wrote in his diary, 'I was desirous of sparing them but could not succeed as it is the will of God that this calamity should befall this city.' Somehow, this logic isn't very convincing, is it?

After Taimur's attack, Vijayanagar, in the far south, became the most important city in India for the next one and a half centuries. We'll talk about this later. Meanwhile, the Tughlaqs were followed by other minor dynasties. Much of their empire was gone but Delhi's urban habitations continued to enjoy political and economic importance. Like every dynasty that has ruled this city before and after, the rulers of this period also built grand memorials to themselves—the Lodhi Gardens is one such example. Now, the rich and powerful of modern India come here for their walks and to discuss world affairs. It is also a good place for birdwatching.

In 1526, a Turko-Mongol adventurer called Babur defeated the Sultan of Delhi in what we call the First Battle of Panipat. We know a lot about Babur because he kept a fascinating diary called the *Tuzuk-i-Baburi*, written in Turkish.

Babur was a direct descendent of Ghengis Khan from his mother's side and Taimur the Lame on his father's side. However, Taimur's empire had been largely lost by the time Babur was born. At the age of twelve, Babur inherited a tiny kingdom in the beautiful Ferghana valley in Central Asia. It could barely support an army of three to four thousand men but even with this small military force, he tried to capture Taimur's capital of Samarkhand many times. He even managed to do it for a brief while but couldn't hold on to it. The Uzbeks chased him out of there and he made his way south with a small band of followers. He won and lost many battles along the way till he gained control of Kabul. And then, he began to look towards India.

It was a daring ambition because his army was much smaller than the Sultan's, but Babur had a secret weapon—

matchlock guns. This was the first time that guns would be used in North India. Babur defeated the Sultan and quickly went on to beat all other rivals, including the Rajputs. And thus began the Mughal (i.e. Mongol) empire in India. The dynasty did not call itself the Mughals though. The name they preferred was 'Gurkhani'—which comes from 'Gurkhan' or 'son-in-law'. Taimur liked to call himself by this name after he was married to a princess from the Ghengiz Khan dynasty.

Ghengiz or Chengiz Khan was the founder of the Mongol empire, one of the largest in history. He united the nomadic Mongol tribes and carried out a number of brutal invasions. The Mongols would eventually come to control lands from Eastern Europe and Iran, across Central Asia, to China.

Although Babur had finally conquered this region, what he really wanted was Samarkhand. He did not think very highly of India. In his opinion, Hindustan 'is a place of little charm. There is no beauty in its people, no graceful social intercourse, no poetic talent or understanding, no etiquette, nobility or manliness. The arts and crafts have no harmony or symmetry. There are no good horses, meat, grapes, melons, or other fruits. There is no ice, cold water, good food or bread in the markets.' Then why did Babur invade India at all? He's quite honest about it: 'The one nice aspect of Hindustan is that it is a large country with lots of gold and money.'

Babur died less than five years after he came to India. After him came his son Humayun, who started the construction of the next Delhi—Dinpanah. Built just south of Feroze Shah Kotla, along the river Yamuna, it included a citadel that we now know as Purana Qila (or Old Fort). As we have discussed, this is said to be the site of ancient Indraprastha but there is nothing to suggest that this is why Humayun chose this place. In addition to the citadel, there was a fortified city. Not much has survived of Dinpanah except one of its impressive gates—the Lal Darwaza (Red Gate). You can see this massive structure across the road from Purana Qila and Delhi Zoo.

But Humayun did not complete either Dinpanah or Purana Qila. He was chased out by a group of Afghan rebels led by Sher Shah Suri. He escaped with his family to Persia and it was Sher Shah Suri who completed the construction of Purana Qila. Though he ruled for a short time, Sher Shah Suri introduced many vital changes. He reorganized tax collection, minted the first silver Rupiya (the earliest version of the modern Indian rupee) and revived the ancient city of Pataliputra (Patna). He also rebuilt the ancient Uttara Path highway from Punjab to Bengal. Known as Sadak-e-Azam (or Great Road), it became a major artery of the Mughal period. The British called it the Grand Trunk Road which, as we know, is now part of the Golden Quadrilateral.

Sher Shah Suri died in a gunpowder-related accident just after five years on the throne. Humayun came back and reoccupied Delhi. But it seemed as if Sher Shah Suri's bad luck had followed him, too. On one fateful day, Humayun went to watch the rise of Venus from the roof of his library. On his way down the steep stairs, he tripped on his robe and died from the fall. Humayun's library building is still there in Purana Qila. The stairway also exists but is not open to the public.

With Humyaun's death, thirteen-year-old Akbar became the new ruler. He is usually called the third Mughal Emperor but it was actually he who created the foundations of a stable empire. Apart from continuing with the changes introduced by Sher Shah Suri, he attempted to improve relationships with the Hindus living in his empire in the second half of his reign. This was a huge step forward.

From the writings of Muslim writers of that time as well as from the ways in which the Hindus responded to the Turkish invaders, it is clear that Delhi's rulers before Akbar regarded themselves as foreign occupiers. Thus, Hindu rulers, especially the Rajputs, saw themselves in perpetual conflict with the Delhi sultans. For example, the Rajput rulers of Mewar did not just see themselves just as kings but as the custodians or guardians of the Hindu civilization embodied in the temple of Eklingji. The deity Shiva was considered to be the real king of Mewar,which is why the rulers did not call themselves 'Maharaja' or Great King. They called themselves Rana which means 'Custodian' or 'Prime Minister'. Mewar suffered huge losses and faced extreme hardship but its rulers still did not give up their fight against the sultans. On three different occasions, its capital Chittaur was defended to the last man and even after the capital fell, the fight continued in the hills.

The shrine of Eklingji is less than an hour's drive from Udaipur. It is a thousand-year-old temple complex wedged into a hillside. The complex is heavily fortified. The fortifications of Chittaur, Kumbhagarh and even Udaipur are within a few hours' drive. Mewar must have been the most militarized place in the medieval world. This was a population that was willing to fight to the death.

By Akbar's time, however, everyone was tired of the fighting that had been going on for centuries. Thinkers like Guru Nanak, the founder of the Sikh tradition, had already argued that both civilizations must learn to accommodate their differences. Emperor Akbar was probably liberal by nature but his thoughts also evolved with time. The change may have happened with the siege of Chittaur, the capital of Mewar in 1568. The fort fell after many months. Almost every soldier was killed. The women killed themselves in a ritual suicide. Akbar further killed 20,000 people—unarmed civilians. Like Ashoka, eighteen centuries earlier, he may have been shocked by the bloodshed he had created.

After this, Akbar began to think about reconciling the two cultures. In 1555, the Mughal nobility or Omrah consisted of fifty-one foreign-born Muslims (Uzbeks, Persians, Turks, Afghans). By 1580, there were 222 in this group but it included forty-three Rajputs and a similar number of Muslims. Not everyone was happy about this. On the one hand, the orthodox Muslims thought Akbar was being too liberal. The rulers of Mewar were still suspicious of his motives and kept up their resistance. The ballads of how Rana Pratap and his army of Bhil tribesmen fought the Mughals can still be heard in the Aravallis of Mewar. His coat of armour and that of his horse Chetak are prominently displayed in the Udaipur City Palace Museum. It took more than a generation before Mewar accepted a more friendly relationship with the Mughals.

Akbar shifted the capital south from Delhi to Agra and then to a newly built city called Fatehpur Sikri. This city took fifteen years to build but was abandoned after only fourteen years because of water scarcity—just like Tughlaqabad. The capital was moved back to Agra. Meanwhile, Delhi remained an important city but it was only a century later, after Akbar's grandson built Shahjehanabad (Old Delhi), that it would regain its full glory.

Akbar did build something grand in Delhi—his father's tomb. The architecture of Humayun's Tomb is not as well-known as that of the Taj Mahal but it is an impressive building by any standard.

It's not that medieval India was only about the building, abandoning, destruction and rebuilding of cities. Most of the population lived in rural areas. Babur tells us that Indian

MEDIEVAL DELHI

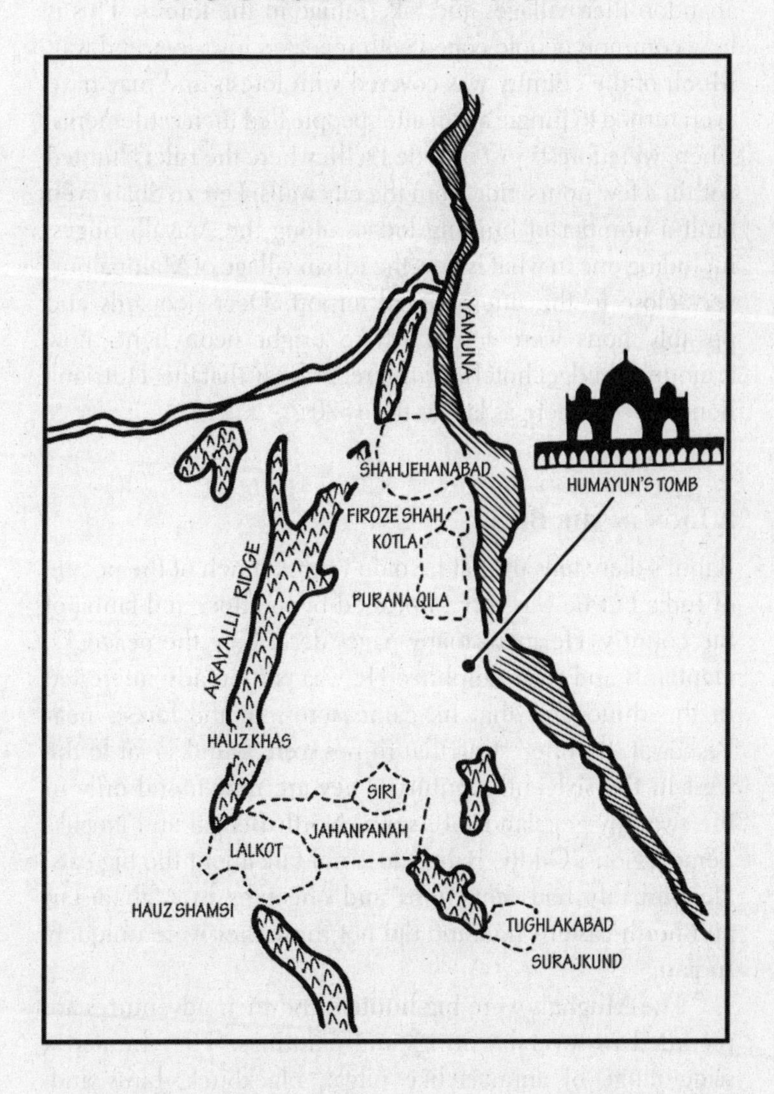

villagers rarely spent any effort in either irrigation or in building permanent homes. Instead, they were always prepared to abandon their villages and take refuge in the forests. This is how common people coped with repeated invasions and war. Much of the country was covered with forests and may have even turned to jungle again after people fled their settlements. There were forests just outside Delhi where the rulers hunted within a few hours' ride from the city walls. Feroze Shah even built a number of hunting lodges along the Aravalli ridges, including one in what is now the urban village of Mahipalpur, very close to the international airport. Deer, leopards and possibly lions were found where bright neon lights now announce budget hotels. British records say that the 'Hurriana lion' was seen here as late as the 1820s.

A Lion in the Bush

Babur's diary tells us that he didn't think much of the people of India but he was very impressed by the flora and fauna of the country. He spent many pages describing the peacocks, elephants and river dolphins. He was particularly interested in the rhinoceros that he came across in the forests near Peshawar. It's interesting that rhinos were found so far to the west in the sixteenth century. They are now found only in the swampy grasslands of Assam, North Bengal and Nepal's Terai regions. Oddly, Babur does not talk about the big cats. He probably had seen lions and cheetahs in Afghanistan and north-eastern Iran and did not think they were uniquely Indian.

The Mughals were big hunters and their adventures are recorded in several writings and paintings. They hunted a wide range of animals like nilgai, blackbuck, birds and, of course, lions. There are relatively fewer accounts and paintings related to hunting tigers. This may be because the Mughals mostly did their hunting in the north-west of the country which had lions and not tigers. We know that there were important hunting grounds near Agra, Delhi, and Bhatinda in Punjab. Most of these places are now thickly populated and farmed but in those times, there were large expanses of uncultivated land near Delhi and Agra as well as on the road to Lahore. These large areas were available to wildlife and several tracts were meant only for the royal hunt. The lion continued to symbolize the power of the State. Only the king and members of the royal family could hunt the animal. Others had to take special permission.

There are many stories about Akbar's lion hunts. Once, in 1568, Akbar went hunting in the Mewat region near Alwar, south of Gurgaon. A lion was spotted and was quickly killed by a hail of arrows shot by his companions. Akbar was annoyed. He issued an order that if another lion were to show up, he would shoot it himself. Another one did come by and Akbar shot it in the eye with an arrow. The wounded animal charged but Akbar could not get a good shot at it though he'd dismounted from his horse. In his excitement, one of the courtiers shot an arrow which further angered the animal and it mauled the man! In the end, the lion had to be finished off by other courtiers.

Akbar was very fond of hunting and was willing to take personal risks, too. In his younger days, he would hunt on horseback or even on foot. He used a large number of trained cheetahs to keep him in the chase. Later in life, he kept a stable of a thousand cheetahs! Over time, however, the emperors got used to hunting from sitting on an elephant's back and hunting with guns—far safer and more accurate. In a single hunting expedition to Rupbas near Agra in February–March 1610,

Jehangir, who succeeded Akbar, and his companions killed seven lions, seventy nilgai, fifty-one blackbucks, eighty-two other animals, 129 birds and 1023 fish—all within fifty-six days!

On a hunt in 1610 in Bari near Agra, one of Emperor Jehangir's courtiers, Anup Rai, came across a half-eaten cow and traced a lion to a thicket. The animal was surrounded and Jehangir was informed. He rushed to the spot and tried to shoot the animal—and missed! The lion charged and the emperor's followers panicked. They ran helter-skelter, even trampling the emperor in their hurry. Anup Rai saved Jehangir's life by battling the lion to the ground with his bare hands till Prince Khurram killed it with a sword. A Hindu Rajput was not just allowed to accompany the royal family on a hunt but was willing to risk his life for a Muslim king—Taimur's descendant, no less. This shows how much the relations between Hindus and Muslims had improved after Akbar. Jehangir gave Anup Rai the title Ani Rai Singhdalan—Commander of Troops and Lion Crusher.

A few years after this, the first English ambassador arrived at the Mughal court. Sir Thomas Roe was a distinguished diplomat and was in India from 1615 to 1619. He became a close friend of Jehangir's. However, this did not mean that he could kill lions freely. In 1617, a lion and a wolf made nightly raids on Roe's camp near Mandu and killed a number of his sheep and goats. He was not allowed to hurt the animals and he had to ask Jehangir for special permission. The permission was eventually given but the lion escaped. The wolf was not as lucky.

Roe says that the symbol of the lion was very important to the royals. One of the royal standards had a lion and the rising sun. The Shahs of Iran as well as the Hindu tradition in the subcontinent shared this love for the lion symbol. The Mughals knew that they were inheriting an ancient imperial dream—Emperor Jehangir inserted his own inscription in Persian on the Mauryan pillar in Allahabad. Thus, the column has inscriptions by three of India's most powerful emperors—Ashoka, Samudragupta and Jehangir—a continuity across eighteen centuries! Why? Whether you take Jehangir's inscription to the Mauryan pillar or the effort to link Akbar to Kalhana's history of Kashmir, the Mughals were trying to build the foundations of their empire in India—as part of the Indian civilization, not an intrusion of it. This was a big shift from how earlier Delhi sultans saw themselves.

The Arabs Rule the Seas

While all these invasions were going on, trade in the Indian Ocean continued to flourish. Marco Polo as well as Ibn Batuta had talked about this. However, the role of the Indians in the trade began to change from the end of the twelfth century. Indian merchants had once been explorers and risk-takers who criss-crossed the oceans in their stitched ships. They could be found in large numbers in ports from the Persian Gulf to China. Buddhist and Brahmin scholars sailed in large numbers to South East Asia where people welcomed them eagerly. But suddenly, a little over a century after the Chola naval raids on Srivijaya, they almost disappeared. What happened?

Around this time, the caste system in India became more rigid and there was a rule that prohibited people from crossing the seas. But this was only a reflection of something larger. There seems to have been a shift in India's attitude towards risk-taking and innovation—a closing of the mind. There are many signs of this that we can see in the culture and civilization of that time. Sanskrit, which used to be a language that changed and grew with the times, stopped taking in new words and usages. Sanskrit literature became obsessed with 'purity'. Scientific progress came to a halt as people began focusing more on learned knowledge rather than experimentation.

Al-Biruni, who visited India around the same time that Mahmud Ghazni was making his infamous raids, commented that Indian scholars were so full of themselves that they were unwilling to learn anything from the rest of the world. He also contrasted this attitude with that of the earlier generation of Indians.

Given these changes, Indian merchants chose to remain on the shores while shipping was taken over by the Arabs. There were also Jews, Persians and Chinese traders. Ibn Batuta noticed a number of Chinese ships in Calicut (Kozhikode) and he describes a military junk that must have been accompanied by a merchant fleet. It was large enough to accommodate a thousand men, six hundred sailors and four hundred men-at-arms. It's clear that Ibn Batuta was looking at a very active trading network on his journeys. There was so much going and coming in this network that he once met a man from Ceuta, a city very close to his hometown of Tangier, first in Delhi and then again in China. He may have been the one to write about his travels but the routes he took were well known to Arab merchants.

The spirit of ancient India was kept alive for several more centuries by the kingdoms of South East Asia. Angkor remained the capital of the Khmer Empire till it was sacked by the Thais in 1431. Its ruins must be seen to be believed—especially Angkor Wat, the largest Hindu temple in the world. The remains of the kingdom of Champa in Vietnam are equally interesting. The kingdom flourished till its capital Vijaya was sacked by Viet troops in 1471. A smaller Cham kingdom limped along till it, too, was overrun in the late seventeenth century. Hindu temples built by the Chams are still scattered across Vietnam—a few are used by the tiny Balamon Cham community that continues to practise Hinduism (they number around 30,000). Sadly, the most important cluster of Cham temples in My Son was heavily bombed by the Americans during the Vietnam War. The site is now a UNESCO World Heritage Site but there is very little to see. The small museum at the entrance has some pre-war photographs that provide an idea about how glorious it used to be once.

Java was another important hub of Indian culture. In the fourteenth century, the Majapahits of Java had established direct or indirect control over much of what is now Indonesia. As they expanded, they pushed out the ancient Srivijaya kingdom based in Sumatra. A naval raid on the Srivijaya capital of Palembang in 1377 finally got rid of their main rival in the region.

One of the Srivijaya princes, Sang Nila Utama, is said to have escaped to the Riau cluster of islands, just south of the Malay peninsula to escape from the Majapahits. One day, so the story goes, he had gone hunting on the island of Temasek where he is said to have seen a lion. So, when he built a settlement here, he named is Singapura or Lion City. And that's how Singapore got its name! However, the animal the prince had seen must have been a Malayan tiger and not a lion. The last wild tiger of Singapore was killed in the 1930s in the neighbourhood of Choa Chu Kang.

One of Sang Nila Utama's successors, Parmeswara, seems to have found it difficult to stay in Singapore because of local rivalries and the continued threat from the Javans. He moved farther north and established his headquarters at Melaka (or Malacca). This is what South East Asia looked like when Admiral Zheng He arrived with the Chinese 'treasure fleet'.

Zheng He was a Muslim eunuch from Yunan. He had been brought as a boy prisoner to the Ming court and castrated. He went on to lead seven major naval expeditions between 1405 and 1433 that visited South East Asia, India, Sri Lanka, Arabia and East Africa. The 'treasure-fleets' were quite remarkable.

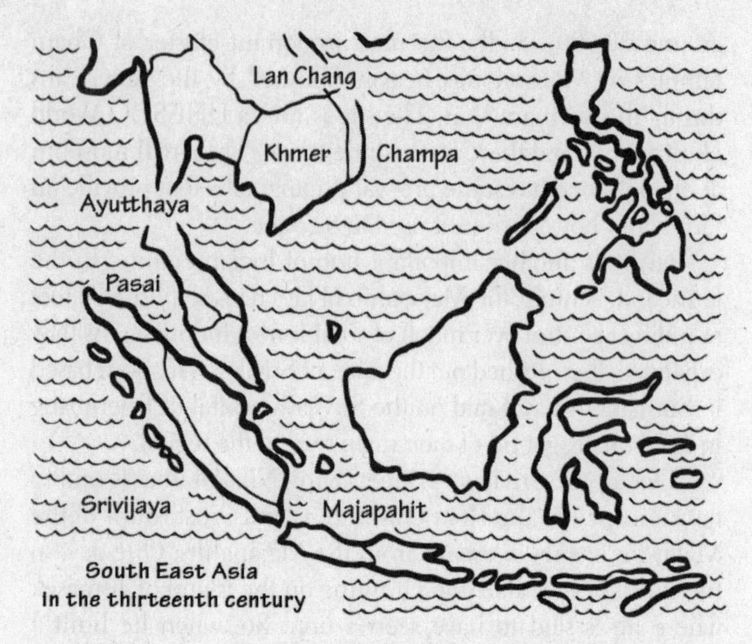

South East Asia in the thirteenth century

There were over a hundred ships and tens of thousands of men. Chinese naval technology at this stage was centuries ahead of the rest of the world. In recent years, some authors have argued that Zheng He may have even visited the Americas. He had the technology to undertake such a voyage but it's unlikely that he actually made the journey across the Pacific.

These naval expeditions were taken for many reasons, including trade and exploration. However, the main goal was to show how powerful China was. The Chinese had defeated the Mongols just a few decades earlier and they were keen to flaunt their importance to the rest of the world. If the sheer size of the fleet did not impress the locals, Zheng He was willing to take military action as he did in a civil war in Sri Lanka.

Around this time, there were already a number of Chinese settlements in South East Asia but the Majapahit Empire of Java was the most powerful. A century earlier, they had defeated the Chinese and the Mongols who tried to control the region. In 1378, the Ming emperor tried to install his candidate on the Srivijaya throne. The Majapahit, annoyed by this interference, simply killed the envoys sent by the emperor.

Zheng He would have been aware of this history. He had a plan to neutralize the Majapahit—supporting Parameswara's new kingdom in Melaka. A large Chinese settlement was created in Melaka and Parameswara personally visited the Ming court. The Chinese encouraged the Melakkans to convert to Islam. Zheng He and many of his commanders were Muslims but he probably did this for political reasons too—the Javans he was trying to defeat were Hindus and it is even possible that the Chinese wanted to reduce the risk of Indians once again influencing this region. The Chinese of this period were very conscious of themselves as a nation and wanted to impress everyone with their power. The Chinese strategy led to the steady Islamization of South East Asia. Melaka boomed while Majapahit slowly lost its powers and the last of the Majapahit princes escaped to the small island of Bali where their descendants continue to live and practise Hinduism. The network of Chinese merchants survived European colonization and they are still an important part of business in the region.

The Chinese domination of the seas, however, came to a sudden end. The mandarin officials in Beijing decided that the voyages were too expensive and not worth it. The treasure fleets rotted and their records were suppressed. Like India, China also closed its mind. Technological superiority could not save them from the change in attitude. For a while, it seemed as if the Indian Ocean would be under the control of the Arabs once again but this was not to be. In December 1497, a small Portuguese fleet rounded the Cape of Good Hope and sailed boldly into the Indian Ocean.

6
Where One-eyed Giants Roam

As we saw in Chapter 3, the people of the Iron Age knew a lot more about geography than the people of the Vedic Bronze Age. The epics clearly show that people knew about the far corners of the subcontinent. They had a fairly detailed knowledge of the terrain by the time of the Mauryan Empire. But, did ancient Indians try to create a map of the country? Over centuries of maritime trade, they would have come to know quite a lot about the geography of the Indian Ocean rim, even as far as the Chinese coast in the west and of the Red Sea in the east. At the same time, they were also quite used to expressing ideas in the form of diagrams, including architectural plans. Because of Aryabhatta, by the time of the Guptas, Indians knew that the world was spherical and even had a fairly accurate estimate of its circumference.

Everything required for map-making or cartography was right there. We would expect that the Indians would have put together all this knowledge and mapped their country and the surrounding oceans. We would also think that they would have created maritime manuals just as the Greeks did (remember *Periplus*?) to help merchants and seamen.

But there is nothing to show that ancient Indians ever attempted to map their country or write down what they knew about the geography of those times. A seaman's manual written in the Kutchi dialect of Gujarat has been found but it exists as a relatively modern copy and nothing is known of its history. It is possible that such things existed but that they have been lost.

This doesn't mean that ancient Indians didn't have a sense of geography. If anything, they were very aware of the layout of the subcontinent and, given their maritime activities, of the Indian Ocean rim. For example, when the famous eighth-century philosopher Adi Shankaracharya set up four monasteries, he chose sites in the four corners of the country—Puri in the east, Dwarka in the west, Sringeri in the south and Joshimath in the north. This is obviously not by chance. But it is probably true that map-making as a science was not very popular in ancient India.

In contrast, the Arabs wrote several books on geography during the medieval period. They also preserved some of the works of classical scholars like Ptolemy of Alexandria at a time when Christian Europe was refusing to accept this knowledge because they thought it was 'pagan'. In the twelfth century, the famous Moorish geographer, Al Idrisi, drew a map that combined his own knowledge with that of Ptolemy. It showed the Indian Ocean as landlocked, an idea that suited the Arabs because this would discourage the Europeans from finding a sea route to the east.

By the fourteenth century, the Persians were drawing maps that show the Indian Peninsula. But the quality of map-making was quite basic. The real experts in this field were the Chinese. They'd been drawing maps of their own country for a long while. By the time of Admiral Zheng He, or perhaps because of his voyages, they had good strip maps of shipping routes through South East Asia and even parts of the East African coast. They are mostly in the nature of sailing instructions rather than accurate physical geography but they are quite detailed and advanced, compared to what the others had.

In the meantime, Europe didn't know anything about the geography of Asia. The Arabs seem to have made sure of that! With the works of classical geographers lost and memories of pre-Islamic trade with India fading, the Europeans didn't have access to proper information. They were also fooled by frauds who exploited their ignorance to make a quick buck.

Sir John Mandeville was an English man who wrote a book full of fantastical tales called *The Travels*. He set off from St Albans in 1322 and returned to England thirty-four years later claiming to have visited India, China, Java, Sumatra and many other places. Geographers, kings and priests studied the book in detail and it was translated into almost every important European language. 300 handwritten copies of the book have survived in various libraries—four times as many as Marco Polo's. According to Mandeville, there were women with dogs' heads, one-eyed giants, geese with two heads, giant snails and other such beings in Asia. He added to the medieval belief that India was ruled by a powerful Christian king called Prester John by giving his own fanciful descriptions of this king and his activities. The Europeans who read this were happy, thinking they may have a Christian ally in the east.

Sir John Mandeville

Funnily enough, these lies had a profound impact on the history of the world. For example, Mandeville was one of the biggest fraudsters of his times but he claimed that his travels had proved that the world was round. This popularized the idea that it was possible to reach India by sailing west. Columbus planned his 1492 expedition after reading *The Travels*, and explorers like Raleigh read the book very carefully. Thus, one of the greatest events of history was based on an elaborate falsehood.

Not all reports by European travellers were fictional. With the sudden expansion of the Mongol Empire in the thirteenth century, the control of the Arabs was finally broken and a few Europeans did travel to the east. The best known of them is Marco Polo. He is today remembered for his travels along the Silk Route to China and his stay at the court of Kublai Khan. However, he returned home by the sea route through South East Asia around 1292. On the way, he visited the ports of southern India and wrote detailed descriptions of what he saw. He tells us of busy ports that exported pepper and imported horses, of Hindu temples and rituals, of diving for pearls, of a royal harem with 500 women, and even of a popular and wise queen who ruled an inland kingdom that produced diamonds—probably the Kakatiya queen Rudrama Devi of Golkonda.

Polo's facts and Mandeville's fiction both fired the European imagination. The fifteenth century was the time of the Renaissance and European scholars opened their minds to the knowledge of classical civilizations. The works of Ptolemy gained importance again and attempts were made to draw maps of India based on his descriptions. The Ptolemaic maps are quite strange—since no maps had survived from the classical times, they were drawn just by reading a text! Therefore, the Ptolemaic maps miss out on basic facts that would have been quite obvious to Ptolemy himself—so much so that he didn't dwell on them in his text. For example, the maps don't show India's coastline as a peninsula but as a long east-west coast. Besides, the texts were over a thousand years old by the time they were used to reconstruct India's geography. And so, they marked places like Taxila although Taxila had disappeared over a millennium earlier! Such was the state of knowledge when the Portuguese decided to look for a way around Africa.

GO, VASCO!

The Portuguese were the first European country to make serious efforts to systematically map the world's oceans. Prince Henry the Navigator, the king's younger brother, became a patron of map-making and exploration. Through the fifteenth century, the Portuguese explored the west coast of Africa and established trading posts and refuelling points. They were quite unhappy when Spain backed Columbus who sailed west based on patchy and wrong information, and yet ended up making one of the greatest 'discoveries' of history! So, they lobbied with the Pope to divide the world into Spanish and Portuguese spheres of influence. As per the Treaty of Tordesillas 1494, Spain was given a claim to all lands west of a meridian of longitude 370 leagues west of the Cape Verde islands. All lands 'discovered' to the east belonged to Portugal.

When it became clear that Columbus had actually not found a westward route to India, the Portuguese were relieved. They had learned from an earlier voyage led by Bartholomeu Dias that Africa could be rounded. A new expedition was prepared in 1497 under Vasco da Gama.

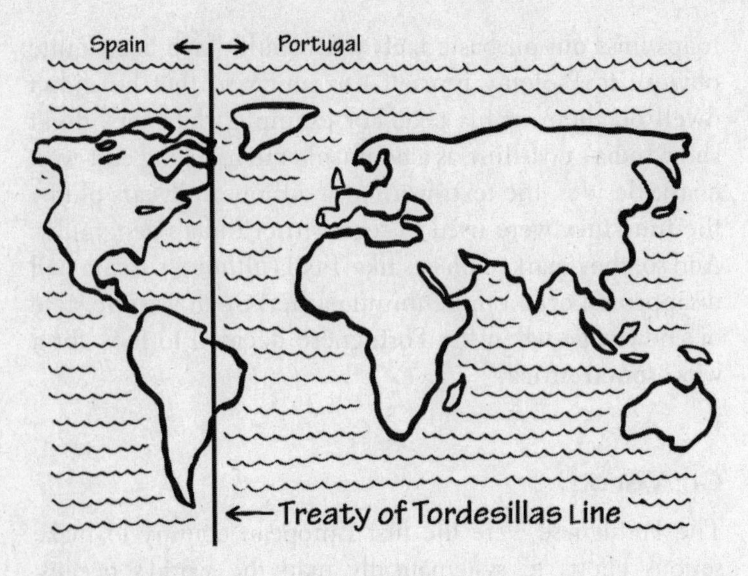

It had three ships—the flagship *San Gabriel*, the smaller *San Rafael* and the traditional *Barrio*. An unarmed supply ship also accompanied these three ships part of the way. The ships would have been quite small and unsophisticated compared to those of Admiral Zheng He, but the Chinese had already withdrawn and were no longer a threat to the Portuguese.

The ships set sail on 8 July 1497 and rounded the Cape of Good Hope by November. After this point, Vasco was in uncharted waters. As they sailed up the Mozambique coast, the Portuguese began to meet Arabic-speaking people. Vasco was quite pleased about this because it proved that he was indeed in the Indian Ocean. The Arabs had already established slaving ports along this coast and some of them had grown into large habitations. However, no one in this part of the world expected the European ships to reach here. The Arabs at first thought that these fair-skinned people were Turks. The Portuguese had the advantage of having several Arabic speakers amongst them as the Iberian Peninsula had only just been liberated from Moorish rule. And so, they were able to talk to the locals and pretend to be fellow-Muslims. When asked by a local sheik for his copy of the *Koran*, Vasco lied saying he'd left it behind in his homeland near Turkey!

This deception could not last forever and they were found out. The Portuguese fended off an attack and quickly sailed farther north in search of Kilwa, an island-city and port that was important enough to be known in Europe. However, they got lost and found themselves in Mombasa which was another port-city. Unfortunately, news of their deception had already made its way up the coast and the Portuguese narrowly escaped being trapped by the sultan of Mombasa.

Vasco da Gama pushed farther north. Along the coast, he made enquiries about Christians and about the kingdom of Prester John. At last he reached the harbour of Malindi, a port that had been visited by the Chinese treasure fleet eighty years earlier and was the source of two giraffes that had been taken back to China. The ruler of Malindi knew who his guests really were but he needed allies against Mombasa and therefore decided to welcome the Portuguese.

Alvaro Velho, one of Da Gama's soldiers, has left us a description of the part-Arab, part-African world of the Swahili coast. The larger port-towns like Mombasa and Malindi had houses built of stone and lime. The population was mainly black African, with a ruling class of Arab origin. The merchants were mainly Arab but some Indians continued to visit these places despite the caste restrictions back home. Remains of this world can still be seen in the Stone Town of Zanzibar, Tanzania. The island of Zanzibar remained a major source of slaves bound for the Middle East till the nineteenth century. It continued to be ruled by an Omani Arab dynasty under British protection, till as recently as 1963.

A community of Indian merchants had visited Zanzibar for a long time but under British protection, many more came to settle there. By the early decades of the twentieth century, there was an active Indian community on the island.

Freddie Mercury, singer of the band Queen, was born here into a Parsi family in 1947. His name at birth was Farrokh Balsara and the house where he spent the first few years of his life still stands in Stone Town.

In a bloody revolution in 1963, however, the Arab dynasty was overthrown, thousands of Arabs and Indians were killed and the island soon became a semi-autonomous province of Tanzania. Still, a small Indian community lives in the narrow lanes of Stone Town, speaking the Kutchi dialect of Gujarat and worshipping in the few temples that still exist.

There is something about Zanzibar that would remind you strongly of the old parts of Kochi and even of old Ahmedabad on the other side of the Indian Ocean. Maybe it's the food, the smell of spices sold in the open, sailing dhows bobbing in the sea or just the weight of centuries of trade with India.

In early 1498, Vasco da Gama had been trying to get a good pilot to guide his ships across the Indian Ocean but he was finding it difficult to find one. In Malindi, he got lucky and the sultan provided him with an experienced pilot described as a 'Moor from Gujarat'. We're not sure about who this pilot was—some say he was the famous Arab navigator Ibn Majid. But we don't really know much other than the fact that he was called Malema Cana.

Both the pilot and the weather proved to be good and the ships reached the Indian coast in just twenty-three days. The open port of Kozhikode (also called Calicut) was filled with vessels of different sizes and the beach was lined with shops and warehouses. Further inland was a large and grand city. The Portuguese ships attracted a lot of attention and the locals rowed up to them, women and children included, to have a closer look.

The ruler of Calicut was Samudrin or Lord of the Sea (mispronounced as Zamorin). He lived in a large palace and was protected by ferocious warriors of the Nair caste. The majority of the people were Hindu, but the Portuguese first thought they were just Christian people who were ill-informed and didn't know any better. Their confusion was probably because of the legends of Prester John and because of the presence of the Syrian Christian community there. The Portuguese corrected their view later. They also noticed that maritime trade was dominated by a large and powerful community of Arab merchants who would not be pleased to see them.

Vasco knew that he had to get back to Lisbon as soon as possible to tell everyone about his findings. The longer he stayed, the greater the danger! The Arabs were likely to trap him or turn the local ruler against him. And so, he went to the Samudri Raja and tried to make the best possible impression with gifts, claiming that he was all for peace. The Arabs did get the Nair guards to briefly hold Da Gama captive but he was soon freed and went back to Europe.

Vasco was given a hero's welcome in Lisbon. King Manuel lost no time in writing to the Spanish monarchs to inform them that the Portuguese half of the world contained India. He also assured them that India was densely populated by Christians. Just that they were not yet strong in faith!

VASCO DA GAMA'S ROUTE

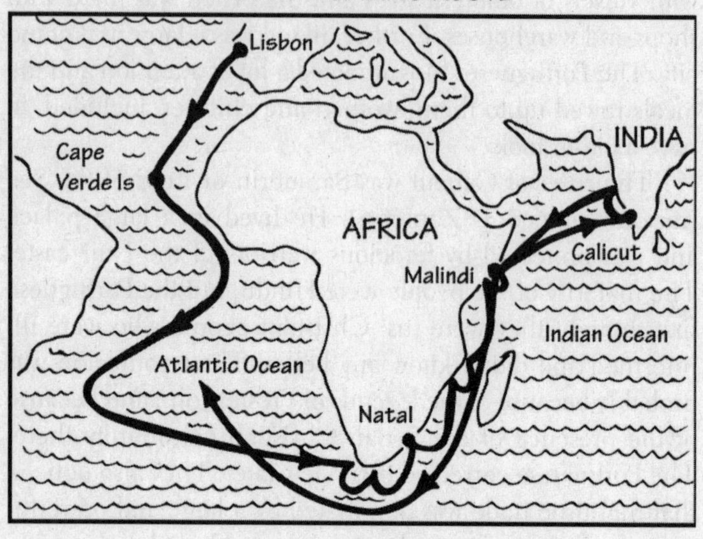

The Portuguese quickly followed up on their discovery. A fleet of thirteen ships and 1200 men were sent under the command of Pedro Alvares Cabral. They were heavily armed with cannon and guns—these were still unknown in the Indian Ocean. By now, the Portuguese had found out that winds and ocean currents made it more efficient to first sail south-west and then turn east rather than hug the coast of Africa. The fleet swung so far west that they landed on the Brazil coast and claimed it for Portugal. Only a small part of Brazil actually fell within the Portuguese sphere as per the Treaty of Tordesillas, but maps were not too accurate then and the Portuguese grabbed more than 'their' share. The Spanish then took the Philippines which was clearly in the Portuguese sphere. Soon, the Treaty came to mean little because other Europeans joined the race.

On reaching Calicut, Cabral presented the Samudri Raja with many lavish gifts before demanding that they throw out the Arabs and trade only with the Portuguese. The king was stunned. While these talks were going on, a Meccan ship loaded with cargo prepared to leave for Aden. The Portuguese seized it, leading to riots in which a number of Portuguese men were killed. Cabral responded by lining up his ship and firing into the city. The Raja had to flee his palace. A number of merchant ships were seized and their sailors were burned alive in full view of the people watching from the shore. And that's how the European domination of the Indian Ocean began. It would last till the middle of the twentieth century.

Within a few decades, the Portuguese used their cannon to establish a string of outposts in the Indian Ocean. Control over Socotra and Muscat allowed them to control the Red Sea and Persian Gulf respectively. In 1510, they conquered Goa and a year later, a fleet sent out from Goa took over Melaka and established control over the key shipping route to the Spice Islands.

Soon, the Portuguese had trading posts in Macau and Nagasaki. They maintained control over the seas with an iron fist and were extremely brutal and cruel in their dealings. They destroyed many Hindu temples and harassed the Syrian Christians for their faith. They did not spare the ships carrying Muslims for the Hajj and sometimes even burnt them mid-sea with the pilgrims on-board.

The Sri Lankans—the Tamils and the Sinhalese—probably suffered the most because of the Portuguese. Much of the island was in a state of almost constant war for one and a half centuries. Thankfully, the Portuguese did not have enough resources to try and conquer the entire subcontinent!

Did you know?

Portuguese control over the Indian Ocean was based on a network of forts along the coast. The best preserved of these forts is in Diu, a small island just off the Gujarat coast. Climbing its ramparts gives you a beautiful view of the Arabian Sea and of the impressive line of sixteenth-and-seventeenth-century cannon. There are few places in the world where you can see and touch such a large number of early cannon, their solid wood wheels bearing the marks left by centuries of rain and sun.

In 1538, the Portuguese were able to defend Diu against a combined attack by the Sultan of Gujarat and a large fleet sent by the Ottoman Turks. A huge Ottoman cannon, cast in 1531 in Egypt, is the only one remaining of this failed Turkish expedition and can still be seen in Junagarh fort. The Portuguese held on to Diu till as recently as 1961. The last of the outposts in Asia, Macau, was handed back to the Chinese in 1999. The Portuguese had been the first Europeans to come to this part of the world and they were the last to leave.

Round World, Flat Map

When Vasco da Gama landed in India, there were around 110 million people living in the country. China had around 103 million, the United Kingdom 3.9 million and Portugal just 1 million. India was still a major economic power with a share of 24.5 per cent of world GDP. But this was smaller than the one-third share of world GDP that it had enjoyed in the first millennium CE. Around 1500, the Chinese economy went past the Indian one in terms of its size for the first time. The per capita income or the income earned per person in India also fell below the global average. After having been behind India for centuries, most European countries had higher per capita incomes.

The richest country in Europe, Italy, had a per capita income that was twice as much as India's. There were rulers like Akbar and Krishnadeva Raya who created periods of prosperity but this did not reverse the trend. The Mughal court was grand and glittering in the sixteenth and seventeenth centuries—but this hid the fact that India was slowly falling behind Europe.

The Europeans were technologically more advanced than everyone else in this period. By the beginning of the sixteenth century, they were simply miles ahead in map-making. The discoveries of the Portuguese were put down on hand-drawn maps and these charts were considered top secret. Before each voyage, the captain was allowed to make a copy from the royal library and was expected to return it with new discoveries marked out when he got back. Not surprisingly, the maps were something the others wanted to steal!

In 1502, Alberto Cantino, an agent of the Duke of Ferrera, stole a chart from Lisbon and took it to Italy. It is preserved in the Biblioteca Estense of Modena and shows that the Portuguese had quickly worked out that India was a peninsula though many elements of Ptolemaic geography were still included.

The first map showing the Indian Peninsula to be published for the public was by Johan Ruysch in Rome in 1508. It shows India as a peninsula and marks a few of the ports on the coast but doesn't show much of the country's interiors. The Indus and the Ganga are the only two Indian rivers marked but their courses are not really accurate. It also shows the Malaya peninsula and marks Melaka. A well-known map by Waldseemüller in 1513 is similar. Over the next century, more maps were published and knowledge about India's geography improved. But mistakes were often made and passed on by map-makers copying information from each other. Empty spaces were filled up with drawings that often looked inspired by Mandeville's fantasies!

However, geographical knowledge went through a major change in the sixteenth century. At the centre of this revolution were Geradus Mercator and Abraham Ortilius, both from the Low Countries.

> The Low Countries form the coastal region in north-western Europe. These include Belgium, the Netherlands, and the low-lying delta of the Rhine, Meuse, Scheldt and Ems rivers, where a large part of the land is at or below sea level.

Mercator was born in 1512 near Antwerp and by the time he was twenty-four, he was already an expert map-maker. He doesn't seem to have travelled to the faraway lands that were being newly discovered but he put together all the information available about them. In 1538, he published his first world map that is one of the earliest to have the names of North and South America. He also showed Asia and America to be separate continents long before the discovery of the Bering Strait proved it.

Mercator lived in a time of religious and political chaos. He was an innovator who asked too many questions and was regarded with suspicion. In 1544, he was arrested for being a heretic, someone who was going against the accepted religious beliefs of the times. Thankfully for Mercator, he had powerful friends, or he was sure to have been tortured or even burnt at the stake! A few years later, Mercator shifted east to Duisburg where he produced his most famous work. In 1569, he produced his world map with the lines 'New and Improved Description of the Lands of the World, amended and intended for the use of navigators.'

The map did not just have better information than earlier maps, it also used a novel way of depicting the

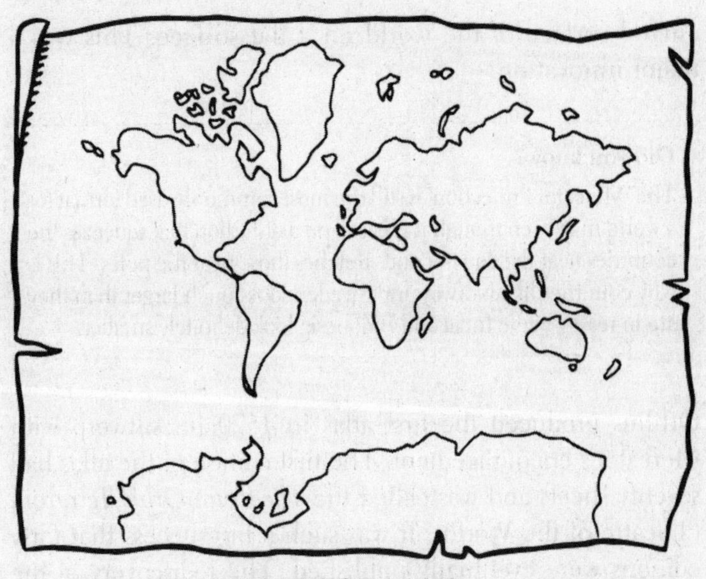

MERCATOR MAP

GALL-PETERS MAP

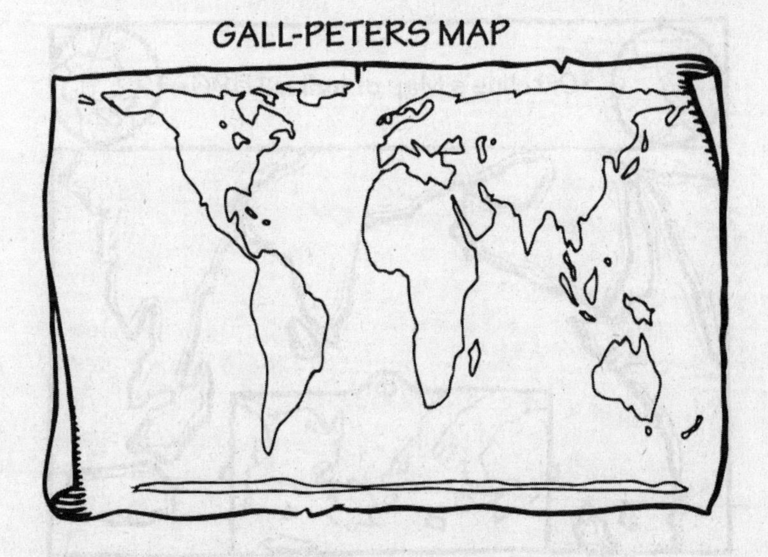

curved surface of the world on a flat surface. This was a major innovation.

Did you know?

The 'Mercator Projection' is still the most commonly used format for a world map even though it is based on a distortion that squeezes the countries near the equator and stretches those near the poles. This is why countries like Norway and Sweden look much larger than they are in reality while India and Indonesia look definitely smaller.

Ortilius produced the first atlas in 1570 in Antwerp with Mercator's encouragement. The first edition of the atlas had seventy sheets and was called the *Theatrum Orbis Terrarum* (Theatre of the World). It was such a big success that forty editions were eventually published. The rediscovery of the classical Greek and Roman works had a deep impact on

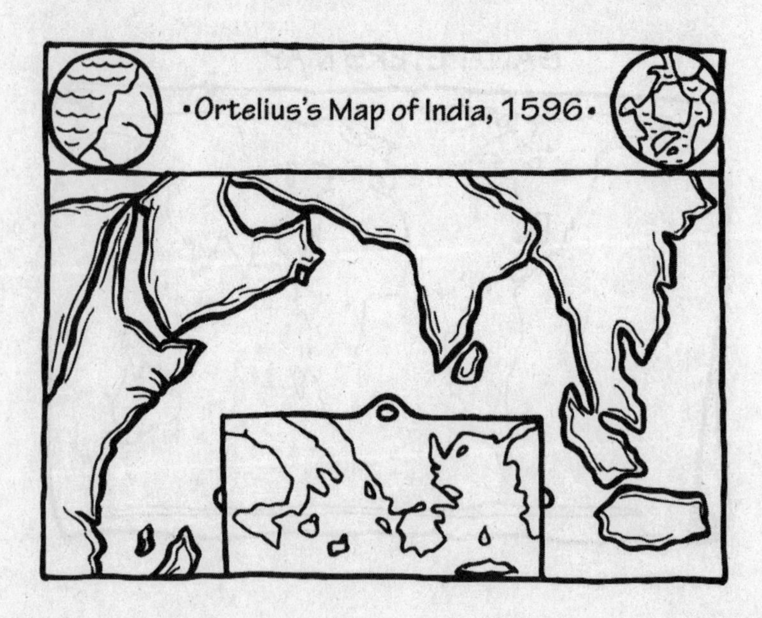

the Europeans of that time. Ortilius took pains to include a map that tried to fit in the new findings about India with the *Periplus* and with Arrian's account of Alexander's expedition. The map locates cities like Pataliputra and Muzaris accurately though so many years had passed.

The City of Victory

A feature very prominently drawn in early European maps of India is the kingdom of Narsinga that covers much of the southern peninsula. Most modern Indians will not know where this place is but it refers to what is now remembered as the Vijayanagar Empire, which was named after its capital city. It was ruled by Narasingha Raya when the Portuguese first arrived in India. He was not a very important king in the history of Vijayanagar but his name stuck and Europeans continued to mark it on their maps long after he and his empire were gone.

The city of Vijayanagar was established just after the brutal raids of Alauddin Khilji's general, Malik Kafur. Around 1336, two brothers, Hukka and Bukka, appear to have got together to defeat rival groups and build a fortified new city. This city was called Vijayanagar or City of Victory. At its height in the early sixteenth century, it was probably the largest city in the world.

The city was built across the river from Kishkindha, site of the monkey-kingdom described in the Ramayana. It is a landscape of rock outcrops and gigantic boulders. This was not a coincidence because the rocky terrain would help to defend the city against the military skill of the Turkic cavalry. An additional advantage was that the place had easy access to iron-ore from the mines of Bellary which were nearby. These mines are still in use today.

A number of visitors have described Vijayanagar in those times, including Abdul Razzaq, envoy from the Persian court, and several Europeans such as Domingo Paes and Fernão Nunes. They say that the city was encircled by a series of concentric walls, as many as seven of them! The largest gap between the first and second walls was used mostly for gardens and farming. Within the inner walls were bazaars, homes, mansions and temples. At the core was a magnificent palace-complex surrounded by strong fortifications. Though the city considered itself to be a place where classical Hinduism was alive and flourishing, it was quite cosmopolitan. That is, it had sizeable numbers of Muslims, Christians and even Jews. Paes tells us that 'the people of this country are countless in number, so much so that I do not want to put it down for fear that it should be thought fabulous.' He goes on to add, 'This is the best provided city in the world . . . the streets and markets are full of laden oxen so much so that you cannot get along for them.'

The remains of Vijayanagar can be visited at Hampi in Karnataka and are simply spectacular. Perhaps only the ruins of Angkor in Cambodia can be compared to those of Hampi in terms of sheer scale. It is too large to be explored by foot and you will need a car and a good guide. As described by the travellers, there is still quite a bit of farming that continues within the UNESCO World Heritage Site. People still use the old canals. There are even remains of a system of stone aqueducts that once brought water into the city.

The remains of temples, palaces and bazaars show that the reports of the city's size were not exaggerated. After decades of excavations, much of the site has still not been uncovered. One of the most remarkable remains is that of Ugra Narasingha—a gigantic sculpture of Lord Vishnu as half-lion and half-man (the Egyptian sphinx is the other way round with the head of a man and the body of a lion). As you will remember, the Vijayanagar empire was called Narsinga by the early Europeans and this sculpture fits right in.

The ruins of the Vijayanagar are located right across the Tungabhadra from Kishkindha. Almost five hundred years ago, Domingo Paes crossed the river and wrote, 'People cross to this place in boats which are round like baskets; inside they are made of cane and outside they are covered with leather; they are able to carry fifteen or twenty people and even horses and oxen can use them if necessary but for the most part these animals swim across. Men row them with a sort of paddle, and the boats are always turning around.' Just like they do now.

In 1565, Vijayanagar was attacked by an alliance of all the Muslim kings of the Deccan. After they were defeated in the Battle of Talikota on 26 January, the Vijayanagar army withdrew instead of defending the capital. The great city was plundered for six months. It never recovered from this attack. Vijayanagar can be considered the last flash of the classical phase of Hindu civilization. The second cycle of India's urbanization had begun on the banks of the Ganga but ended on the banks of the Tungabhadra.

The King of the World

By the late 1500s, the Portuguese and the Spanish had competition from rival European nations. In the autumn of 1580, Francis Drake returned to London after he had gone around the world. By 1588, the English had decisively defeated the Spanish. However, it was the Dutch who first took on the Portuguese in the Indian Ocean. The Dutch, only recently free from Iberian rule, set up the United East India Company in 1602. In the following year, they had a trading post in Banten, West Java and by 1611 in Jayakarta (later Batavia and now Jakarta). Soon they were challenging the Portuguese along the Indian coast and in Sri Lanka.

Apart from this, the Dutch also had better maps. Thanks to Mercator and Ortilius, they had the upper hand. A map of the Bay of Bengal by Janssan and Hondius printed in the 1630s captures the improvements in the level of knowledge since Waldseemüller a century earlier. The map shows Sri Lanka, the eastern coastline of India, Bengal, the Burmese coast, the Andaman and Nicobar Islands and the northern tip of Sumatra.

A lot of details along the coast are shown with major and minor habitations marked out. These include Masulipatnam and Pallecatta. The temple town of Puri in Orissa is marked as Pagod Jagernaten after the temple to Lord Jagannath. Since it is a chart for ships, depth measurements are marked out in a number of places such as the Gangetic delta. For the first time, we have some concrete information of the interior of the country. For example, the riverport of Ougely (Hooghly) is clearly pointed out. Hooghly was the most important trading centre in eastern India and even though it's not as important now, this channel of the Ganga is now named after the old port.

In the meantime, the English had also formed their own East India Company. By 1612, they had set up their first factory at Surat, Gujarat. The company's position became stronger because of the embassy of Sir Thomas Roe to the court of Emperor Jehangir. Roe presented an atlas of the latest European maps to the Mughal court but these maps were politely returned after four days. Was it because the courtiers didn't understand the maps? It could also be that the maps showed how tiny the Mughal Empire was compared to the world known to the Europeans. Maps have always been about politics, just as they are about geography. Even now, India and China are fighting about how Arunachal Pradesh is shown on the map.

Shah Jehan came to power after Jehangir in 1628. The name Shah Jehan means 'King of the World' in Persian and his rule was the golden age of Mughal architecture. Many monuments, small and great, including the Taj Mahal, were built under him. He also decided to move the capital back to Delhi and build a new city in 1639. The city was called Shahjehanabad—what we now know as Old Delhi.

Shahjehanabad was completed in 1648. It had twenty-seven towers, eleven gates and a population of around 4,00,000. Shah Jehan had chosen a place that was north of the existing city, the northernmost Delhi built so far. It contained a walled palace-complex surrounded by walls of Red Dholpur sandstone—what we call the Red Fort. For lesser buildings, material was taken from older Delhis, especially Dinpanah and Feroze Shah Kotla. The Red Fort was built along the river and during the monsoon, water would have flowed along the palace walls. However, most of the time, there was a beach between the river's edge and the fort where elephant-fights and other events were organized for the entertainment of the court.

Old Delhi has gone through many changes since these times but you can still see some features that continue to exist. There was a straight and wide avenue that began at the Red Fort's western gate and ran through the main bazaar to one of the city's main gates—Chandni Chowk! It was named after the way the full moon once reflected on a canal that ran along the middle of the road.

The French traveller Bernier visited the city a few decades after it was completed. One of the first things that struck him was that the fortifications of both the city and the

Red Fort were old-fashioned and not designed to withstand a military attack. Why did Shah Jehan go for such outdated designs? Was it because he felt that his empire was safe and that nobody would try and attack it? Or does it show us that India was already technologically behind the West? Whatever the reason was, this would prove to be a major problem as the Shahjehanabad walls repeatedly failed to hold off attackers over the next two centuries.

Bernier describes the grand palaces of the nobility with their courtyards and walled gardens. He tells us that the rich had raised pavilions set in the middle of gardens and open on all sides to allow the breeze to flow from any direction. The insides of the private apartments had cotton mattresses covered in cloth in summer and carpets in winter. Cushions of brocade, velvet and satin were scattered around the rooms for the use of those sitting down. All of this can be seen in Mughal paintings and buildings that have survived from that time across northern India.

Delhi was not a city that just had grand palaces and mosques. The majority of the people were common folk—shopkeepers, artisans, servants, soldiers and so on. These people lived in huts made of mud and straw that were built between and around the great palaces of the nobility. This means that Shahjehanabad had many slums—just as modern Indian cities do. These slums made it look like the city was a collection of many villages.

Fires were common and Bernier reports that sixty thousand roofs had been gutted in just one year—this might have been an exaggeration but it shows how big a problem this was. This issue had been described 1800 years earlier by the Greek ambassador Megasthenes when he visited Mauryan

Pataliputra but it was still not solved in Mughal India. Mughal Delhi was a city of extremes. In Bernier's words, 'A man must either be of the highest rank or live miserably.'

The Frenchman describes the bazaars as busy, chaotic and dirty—not very unlike Old Delhi now! He says there were many *halwais* all over the city but there were many flies too. And dust. There were also shops selling a variety of kebabs and meat preparations. Old Delhi still has these. You get off at the Chawri Bazar Metro stop and then take a rickshaw to the Jama Masjid. Go late at night when the lanes with the kebab-and-sweet-shops are full of people. With smoke rising from the open ovens and the old mosque in the background, it looks so much like medieval Delhi must have!

> Bernier was not very tempted by the kebab shops. He says, 'There is no trusting their dishes, composed for aught I know, of the flesh of camels, horses, or perhaps oxen who have died of disease.' Did Bernier suffer a bad case of Delhi Belly during his stay?

Peacock Kebabs, Anyone?

At the time that Bernier was travelling through the Mughal Empire, there were many other Europeans—merchants, officials, mercenaries, adventurers—who were also in the country and have left us many colourful accounts of their experiences. One of these was Jean-Baptiste Tavernier, also a Frenchman.

In the winter of 1665–66, Tavernier travelled from Agra to Bengal and wrote about his experiences. He says that the imperial highways were full of bullock-cart caravans carrying rice, salt, corn and so on. Although most caravans were made of one or two hundred carts, some were huge with 10,000 to 12,000 oxen! Remember the traffic jams we talked about earlier?

The bullock-cart caravans were driven by nomadic castes called Manaris who travelled the trade routes with their families and belongings. At every stop, they would set up their tents and make a temporary village. Each group had a chieftain who usually wore a string of pearls. There were quarrels between the leaders of rival caravans and sometimes, these got so intense that the disputing parties had to be taken to the emperor! So the ill-tempered truck drivers who now drive on these highways have a long history!

Tavernier says that the travellers had a choice between two kinds of transport—light carriages pulled by bullocks and palanquins carried by men. The carriages cost about a rupee a day and came with cushions and curtains. The palanquins needed about six people to lift and a long journey would mean at least twelve people were there so they could take turns. Each man cost four rupees a month. Some people who wanted to show off would hire twenty to thirty armed guards who came with muskets and bows! These cost as much as the palanquin bearers but were higher in status.

You must have seen modern-day VIPs travelling with little flags on their luxury cars. Tavernier says that in those times, the English and Dutch officials insisted on a flag bearer who walked in front of the party in honour of their respective companies.

In addition to the Sadak-e-Azam highway through the Gangetic plains, there were many other internal trade routes that continued to thrive. As per a tradition that went back to ancient times, trees were planted all along the way to provide shade. This custom was followed even in the twentieth century but somehow, we've stopped planting trees along major highways now.

In the south, the road through the Palghat Gap continued to be used to connect the ports of the Kerala coast with the interior. However, the old Dakshina Path seems to have been used lesser and lesser in this period. Instead, there were a number of important trade routes that linked the imperial capitals of Agra and Delhi with the ports of Gujarat. For example, a route used by Peter Mundy of the English East India Company began in Agra and made its way south-west through Fatehpur, Bayana, Ajmer, Jalore, Mehsana, Ahmedabad and finally to Surat. Another route was to travel more directly south from Delhi-Agra to Dholpur, Gwalior, Narwar, Ujjain and finally to Mandu. From Mandu, the route turned west to Surat. Bernier tells us that goods from Surat made it to Delhi in four-six days.

Some of these places were important towns but there were many camping areas along the way though their quality wasn't always the same. The larger ones had spacious walled enclosures where merchants could spend the night safely. Travellers could draw water from the wells and buy provisions. Many of the busy roads had water-stops or 'piyaus'. Giving a thirsty traveller drinking water was supposed to give one *punya* (religious merit) and people built piyaus in memory of their loved ones. Many of these old ones still exist alongside the new ones. The road from Delhi to Gurgaon (MG Road) used to have many old piyaus till recently. The last one was demolished in 2009 to make way for the new Metro line.

The quality of the road and accommodation could sometimes be terrible! The Portuguese Catholic priest Friar Sebastian Manrique wrote down many amusing anecdotes of his travels through Orissa and Bengal during the monsoons of 1640. After leaving Jalesar, the priest and his companions found themselves in a small village which did not have a proper caravanserai. They had to spend the night in a large cowshed! The cows were not the problem though. The travellers were attacked by a swarm of mosquitoes! And then, it began to rain and they discovered that the roof leaked! It was almost dawn when Sebastian Manrique was able to sleep but not for long. The cowshed was suddenly full of birds, including two peacocks! The friar's companions decided to kill and eat the peacocks. They knew that the locals thought the birds to be sacred and tried to hide what they had done. But the truth was discovered by their hosts and an armed mob gathered outside. The friar's party fled, firing muskets. Such was life on the road in seventeenth-century India!

Guerrilla Attack!

By the time Bernier and Tavernier were criss-crossing India, Shah Jehan was no longer the emperor. Aurangzeb, his son, had grabbed the throne after imprisoning his father in Agra fort and ruthlessly killing all his siblings. The new emperor next attempted to expand the boundaries of the Mughal Empire.

Aurangzeb's big push was into the southern peninsula. He shifted to the Deccan in 1682 and would never see Delhi again. He lived in a constant state of campaigning for the next twenty-six years. Aurangzeb extended the empire but he also destroyed it. The never-ending wars were disastrous—for the land and the exchequer. Bernier commented that though the Mughal emperor had revenues that exceeded the combined ones of the Shah of Persia and the Ottoman Sultan, he was not wealthy because all of it was eaten by the expenses.

Aurangzeb was also a religious bigot, a man who could not tolerate people of other faiths. He destroyed Hindu temples and reimposed the hated jiziya tax on non-Muslims. When this tax was first announced, the Hindus of Delhi gathered in large numbers in front of the Red Fort to protest against it. The emperor set his elephants against them and many were trampled to death. There were many other atrocities that Aurangzeb committed in the name of religion.

Because of all this, the relationship between the Hindus and the Mughals became sour. There were revolts in many places across the empire. One of the most successful of these was led by Shivaji, the Maratha rebel. The exploits of Shivaji and his men are so daring that it would have been hard to believe them if not for the people who wrote about those events in those times. Using the volcanic outcrops of the Deccan Traps (which you read about earlier), the Marathas repeatedly outwitted the larger Mughal armies.

The Marathas captured Sinhagadh by using a trained monitor lizard named Yeshwanti to scale the walls! The guerrillas tied a rope around the lizard, which climbed up a rock face that was so steep that it had been left without any guards. A boy then climbed up the rope and secured it for the rest. The fort of Sinhagad is just outside Pune. You will see cadets from the nearby military training school climbing up the hill with their heavy packs.

Another group that broke out in open revolt were the Bundelas. Their leader, Raja Chhatrasaal, used the low hills of the Vindhya range to wage a campaign against the Mughals. It is said that Raja Chhatrasaal had a very beautiful dancer named Mastani in his court. When the Marathas rescued the Bundela chief from a tight spot, Chhatrasaal thanked the Maratha commander Baji Rao by 'gifting' Mastani to him. Baji Rao went on to become the Peshwa (Prime Minister) of the Marathas. Mastani rode with him on many of his campaigns. On the highway between Orchha and Khajuraho, there is a small but beautiful palace built on a lake by Chhatrasaal for Mastani during her younger days. The surrounding hills are heavily fortified which goes to show how troubled those times were.

But who first defeated the Mughals? It wasn't the Marathas or the Bundelas. This defeat happened in the middle of the Brahmaputra in faraway Assam at the hands of the Ahom general Lachit Borphukan. The Ahoms came to India as refugees in the early thirteenth century. They were distantly related to the Thais from what is now the Burma-China border and were probably just a few thousand in number. Soon, they converted to Hinduism and established a kingdom that lasted from 1228 till 1826.

In 1662, Aurangzeb's governor in Bengal, Mir Jumla, attacked the Ahoms of Assam, but couldn't fully defeat them because of heavy rains, the difficult terrain and the constant guerrilla attacks. This raid hurt the Ahoms but they survived and steadily got back their territory. In 1671, their commander Lachit Borphukan cleverly coaxed the Mughals into a naval battle on the Brahmaputra river where the smaller and more manoeuvrable Assamese boats inflicted a crushing defeat on the Mughals. Though seriously ill, Lachit Borphukan personally led the attack. This was the first major defeat the Mughals had faced in India and their empire began to crumble.

Despite the defeats and rebellions the Mughal Empire survived many things—religious intolerance, leaky public finances, Maratha guerrillas, Bundela chieftains and the Assamese navy. The foundations built by Akbar and those after him were still strong but Aurangzeb committed the ultimate sin—he stayed on the throne too long. He was ninety by the time he died in 1707! Just as it happened with Ashoka and Feroze Shah Tughlaq, those who came after him were weak rulers. This led to a foreign invasion.

In 1739, the Persian army of Nadir Shah occupied Shah Jehan's Delhi and killed twenty thousand people. They left with much treasure, including the famous Peacock Throne. The power of the Mughals declined so much that the Marathas occupied large parts of central India even as governors of far-flung provinces like Bengal and Hyderabad became virtually independent. Eighteenth-century India had descended into chaos!

A number of foreigners saw this as an opportunity. In the North West, the Afghans under Ahmad Shah Durrani, and in the North East, the Burmese set their eyes on India. In the coastal regions, the rivalry between the Dutch and the Portuguese had been replaced by the rivalry between the French and the English. Armies for hire wandered around the countryside, feared by rulers and the common people. For a short while, it looked like the Marathas would replace the Mughals and establish order but their internal rivalries let them down. They were defeated by the Afghans in January 1761 in Panipat, Haryana. The scene was set for a war of maps.

The War of Maps

The Marathas were the only Indians who had developed some map-making ability. Their maps were not as good as the European ones, but they knew their terrain. Meanwhile, the French and the British map-makers became the experts in this technology, replacing the Dutch.

At first, it was the French who held the advantage. By the early eighteenth century, they had a well-established network on the Indian coast. Their most important outposts were in Pondicherry, just south of Madras (Chennai) and the ancient submerged port of Mahabalipuram. There were smaller outposts like Mahe on the Kerala coast, Yanam on the Andhra coast and Chandannagar on the Hooghly channel of the Ganga, just north of the English settlement at Calcutta. The French also controlled the strategically important island of Mauritius in the middle of the Indian Ocean.

The maps of India that the French made were far better than those created by their rivals. The best of the French map-makers was D'Anville. He never visited India but seems to have collected the best available information from his Paris home. Unlike others before him, he focused on the facts and avoided the fantastical elements—Mandeville's influence was finally wearing off!

Thus, when D'Anville wanted to correctly locate Satara, the Maratha capital, he asked the Portuguese ambassador to the French court for more information. The Portuguese were fighting the Marathas at that time. D'Anville was told that Satara was in the Ghats and that it was eight days' journey from both Goa and from Bombay, at the apex of a triangle formed by these two lines and the coast. For most map-makers of that time, this would have been more than enough information. But D'Anville was not satisfied. He left Satara out because he felt he couldn't locate it exactly.

The British were not too far behind. There were many British map-makers in the first half of the eighteenth century—Herman Moll, John Thornton and Thomas Jefferys. Their records show that they also followed the developments in French maps. They made detailed local maps of specific ports and military places. One of the more interesting of these maps is an English map of Maratha admiral Kanoji Angre's sea fort. From its fortified base at Vijaydurg, the Maratha navy troubled European shipping up and down the Konkan coast for several decades. Angre also defeated the Abyssinian pirates, the Sidis, but was unable to remove them from their base at Murud-Janjira.

The forts of Vijaydurg and Janjira lie south of Mumbai. The fort of Janjira is built on a small island but local fishermen are happy to take visitors out on a rowboat for a small sum. Vijaydurg is built on a peninsula but also offers spectacular views of the Arabian Sea. The eighteenth-century English map of Angre's fort contains a lot of details about its defences.

It also shows what the Europeans thought of the Maratha admiral. They've marked a building as 'Godowns where he keeps his Plunder'—as if he was just a pirate!

Though the maps of India by now showed detailed depth measurements along the coast and even greater detail for the entrances of major ports, they had relatively little idea about the Himalayas! The Himayalas are one of the most prominent geographical features of the planet. Most maps do show some awareness of mountains to the north but the range is not really marked anywhere properly. There was a belief going back to the time of Alexander that the northern mountains were a continuation of the Caucasus.

But Bernier visited Kashmir and he left a comprehensive account of the province which was used by the Mughal emperors as a summer retreat. He says there were two wooden bridges over the Jhelum at Srinagar and beautiful gardens along the riverbanks. Most of the houses were made of wood though some larger buildings, including the ruins of old Hindu temples, were made of stone. He talks of the pleasure boats the rich owned that floated in the Dal Lake and about the lovely parties they threw in the summer.

Bernier says that the Mughals used their base in Kashmir to extend their influence into Little Tibet or Ladakh and Greater Tibet (Tibet itself). Bernier wasn't too impressed with the stunningly beautiful place. He says, 'No other useless place can be compared with it.' But a visit to Ladakh will make you disagree with him for sure! Try spending a full-moon night on one of the lonely mountain passes. It's impossible to describe the way the stars look at these heights and the way the moonlight reflects on the bare rocky mountainsides. The moon can be so bright that you can almost read a book by it!

It looks like the Mughals made some inroads into Ladakh. The Ladakhis promised to pay an annual tribute to them, allow the building of a mosque in their capital and to issue coins in the name of Aurangzeb. The mosque in Leh can still be visited. It's at the head of the main bazaar and just below the old palace.

However, because of the difficult terrain, the Mughals could not really make sure that the Tibetans submitted to their rule. Bernier says that nobody really believed the Tibetans would keep their word. He was very curious to know more about Tibet and the Dalai Lama. Bernier tried to question Tibetan merchants about their country but learned very little. As we shall see, the British had to make great efforts to acquire reliable information about this land in the nineteenth century. For now, the Europeans needed to know more about the geography of the subcontinent itself.

So far, knowledge of India's interiors was quite basic. They only knew of the major trade routes. But this changed with the Battle of Plassey in 1757 where the troops of the English East India Company, led by Robert Clive, defeated Siraj-ud-Daulah, the Nawab of Bengal. For the first time, a European power came to control a major province. Soon, the British acquired large territories and led campaigns to the deep interiors of the country. Accurate maps became more important than ever. Enter Colonel James Rennel.

7
Here Comes the Train

The Portuguese first came to Bengal in 1530. They set up trading posts at Chittagong in the east and Satgaon in the west. Over time, the river near Satgaon silted up and the river port of Hooghly became the main trading hub. The port was on the Bhagirathi, a distributary of the Ganga—what we now call the Hooghly after the old port town.

By the seventeenth century, other Europeans also joined the party. They set up trading posts along the river—the French at Chandannagar, the Danes at Srirampur and the Dutch at Chinsurah. The English East India Company first had its local headquarters at Hooghly. However, they seem to have had some problems with the local Mughal officials and were forced to sail down the river in 1686. When matters finally settled two years later, the English sent a squadron on ships from Madras (now Chennai) to re-establish their presence in Bengal. The squadron was headed by the company's chief agent Job Charnock.

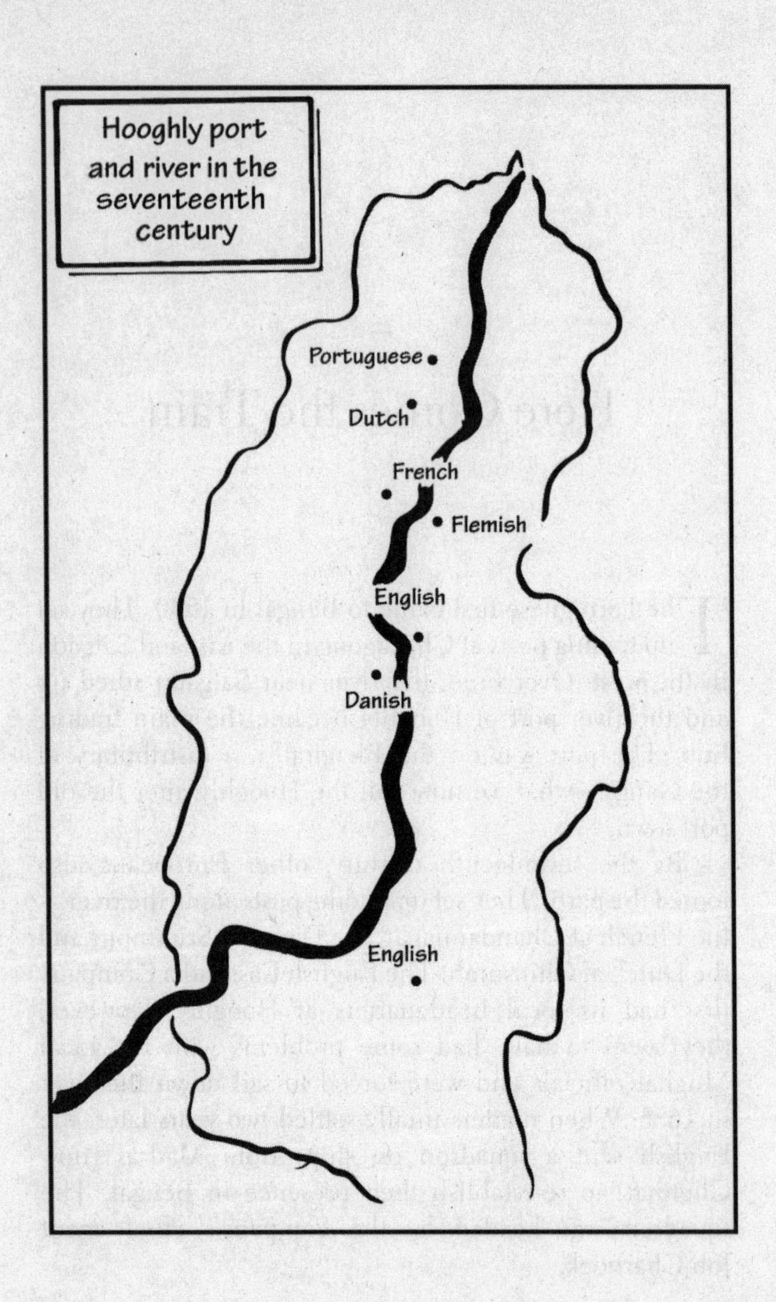
Hooghly port and river in the seventeenth century
Portuguese
Dutch
French
Flemish
English
Danish
English

Calcutta Calling

On 24 August 1690, Charnock landed at a village called Sutanuti on the east bank of the river. He had already visited the spot two years earlier and had liked it. So he decided to build the new English trading post here. It would grow into the city of Calcutta, now called Kolkata.

There were three villages in this area—Sutanuti, Gobindapore and Kalikata. The last village gives the city its name. The merchant families of the Setts and Basaks already ran big businesses here. There was a fourth village nearby, called Chitpur, from where the road ran all the way to the ancient temple of Kalighat. Just off this road, in the middle of a jungle full of tigers, was a Shiva temple built by a hermit named Chowranghi. The temple is no longer there and the place is now occupied by the Asiatic Society on Park Street. But Chowringhee Road, which is one of the city's most important roads, is named after the hermit. The road was renamed after Jawaharlal Nehru in the 1980s but most citizens of Kolkata still call it by its old name.

Job Charnock probably chose this place because of its military advantages. The river ran along the west of the site while there were marshy salt lakes to the east. To the south there were dense jungles full of tigers, while to the north there was a creek that ran from the river to the salt lakes in which big boats could travel. You can still see many of these features. The creek no longer exists but the places surrounding it have names like Creek Row and Creek Lane. The eastern marshlands where the city would expand in the 1970s is still commonly called Salt Lake although its official name is Bidhannagar. A few of the lakes still exist as the East Kolkata Wetlands and these give the city a unique natural sewage recycling system.

The British who first arrived in this area settled down around a water tank called Lal Dighi, which had been excavated by a Bengali merchant Lal Mohan Sett. The name Lal Dighi means Red Pond and there's a story that it gets its name from the colours used by the locals during the festival of Dol (or Holi). The water tank still exists and stands in the middle of the business district. Soon, the British built a number of big buildings around Lal Dighi, including a fort they named Fort William. The General Post Office now stands in the place where the original Fort William used to be and should not be confused with the later Fort William that we see today.

Though trade flourished in this area, there were also many difficulties. The region was surrounded by swamps full of mosquitoes and many early European residents of Calcutta died due to disease. Alexander Hamilton, who lived in Charnock's times, says that there were 1200 English of various ranks living there when he visited the city. Within six months, 460 of them died! This may have been a really bad year but it gives us a sense of what kind of problems the East India Company employees had to encounter. Less than three years after establishing the trading post at Calcutta, Job Charnock also died. His tomb is on the grounds of St John's Church, just off Lal Dighi. His eldest daughter, Mary, passed away a few years later and was buried in the same tomb.

Nevertheless Calcutta continued to grow. A map from 1757 shows that the British had built a fortified trench called the Maratha Ditch all around Calcutta to defend it from attacks by Indian rulers. The name of the ditch tells us that the British saw the Marathas as more of a threat than they did the Mughals after the death of Aurangzeb. Most of the area within the fortifications was still mostly rural but there is a small urban cluster around Lal Dighi and along the river.

In 1756, the Nawab of Bengal Siraj-ud-Daulah briefly took control of Calcutta and renamed it Alinagar. But just a year later, Robert Clive defeated him at Plassey and the British came to control the province. Calcutta now became the headquarters of a rapidly expanding empire. Over the next century, it became the largest city in the subcontinent and one of the most important urban centres in the world. This is clear when you compare the 1757 map of Calcutta with the one published by Chapman and Hall in 1842. A few of the old features are still there. Lal Dighi is shown but is surrounded by large buildings including the Writers' Building. This is not the Writers' Building built in 1882 which functions today as the secretariat of the state of West Bengal. The original Writers' Building was also a big building and was used as rent-free accommodation for clerks and other junior employees of the East India Company. The Maratha Ditch has been filled up but you can still see its outline in the 1842 map as the Upper Circular and Lower Circular roads—they continue to be very important roads even now though they have new names.

If you've visited Kolkata and are familiar with the city, you will find the 1842 map to be very interesting. The form of the modern city is clearly visible. The old Fort William has been replaced by the large star-shaped fort that is still used by the Indian army as its eastern headquarters. The British town planners left large open spaces around the new fort—this was so there would be a clear line of fire for the fort's cannon. These are now the parks of the Maidan. The Victoria Memorial did not exist at this stage and in its place is the complex marked as the Grand Jail. The site of the Turf Club already has a racecourse. Well-known roads such as Park Street and Camac Street have taken shape and are clearly marked. Many of the street names have been changed since the 1970s—otherwise, you could probably find your way around most of central Kolkata by using the 1842 map!

The map also shows how, by the mid-nineteenth century, the fast growing city was spilling out of the limits, the old Maratha Ditch. We can see how the new suburbs of Sealdah,

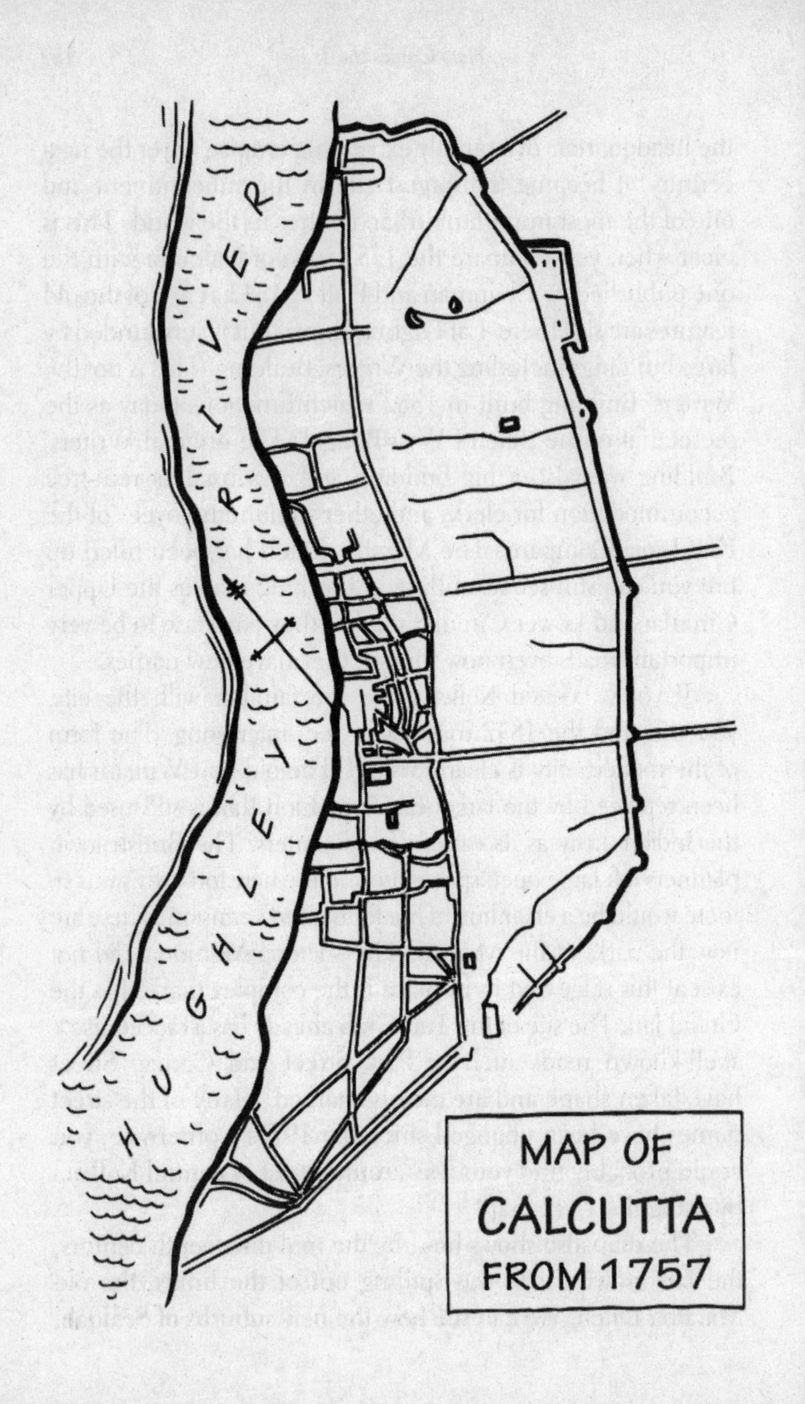
HUGHLEY RIVER
MAP OF
CALCUTTA
FROM 1757

Ballygunge and Bhowanipur are just beginning to appear. They turned into fully urban settlements very slowly. Even in the early 1980s, some parts of Ballygunge still looked semi-rural and had big bungalows, fish ponds and weekly village markets. These open spaces are now full of multistorey residential towers but some reminders of the past are still present—the peculiar lanes, the odd hut standing in between modern buildings, the old village shrine in the middle of the road . . . By the middle of the nineteenth century, Calcutta also became a centre for intellectual and cultural activity. Indians from across the subcontinent came to the city to earn a living. There were large communities of Jews, Armenians, Greeks and even Chinese in the city. Though these communities have reduced in size in recent decades, they have left behind buildings and names of places that still remind us of them. This environment with so many cultures living together set the stage for the next phase of evolution of India's civilization. Over the next century, Calcutta attracted social reformers like Ram Mohun Roy, Swami Vivekananda and Vidyasagar who pushed through remarkable changes that have shaped modern India.

These early social reformers also argued in favour of providing education to Indians in English. This was a choice that went on to have a deep impact. Many think that English education was used by the British to create a class of Indians who would be loyal to them. Thomas Macaulay in 1835 argued, 'We must at present do our best to form a class who may be interpreters between us and the millions whom we govern; a class of persons, Indian in blood and colour, but English in taste, in opinions, in morals, and in intellect.' His note is often used by people to support the view that English education was used only for the purpose of creating loyalty amongst Indians. But not all Britishers agreed with Macaulay. The fact was that many Indian reformers also favoured English—this is not so strange because these reformers knew that Indian civilization had been in decline for a long time. They correctly blamed this on lack of technological and intellectual innovation.

The knowledge of English was regarded as a window to the world of ideas that came from Europe. Far from creating a class of loyal Indians, it was the English-educated middle class that would be at the forefront of India's struggle for independence!

The College of Fort William, which was set up for training British civil servants, was one of the important places for Anglo-Indian interaction. The college was meant for training civil servants but it brought about remarkable interaction between Indian and British scholars. This led to new scholarship as well as thinking. One of these scholars was Ishwarchandra Vidyasagar, who taught there in the 1840s. He was an extraordinary man and his contributions include giving the Bengali language its modern form, the emancipation and education of women and the teaching of Sanskrit texts to low-caste Hindus. Indian civilization benefited a lot from this new way of thinking.

The students of the College of Fort William were not always thinking about their studies. A student named Mr Chisholme was sued in 1802 and brought to court by Jagonnaut Singh, a lawyer. Here is what happened: a cat had been sitting in a shop near Chisholme's residence. The student set his dog on the cat but it fled into the lawyer's house and into the women's quarters! Mr Chisholme and the dog followed the cat. When the lawyer objected, Chisholme punched him in his forehead! In the end Chisholme admitted his guilt and was reported for proper action.

Meanwhile, Thomas Stamford Raffles, a talented young official, was sent by Governor-General Minto to Penang (now in

Malaysia) to keep an eye on the Dutch in South East Asia. The British and the Dutch had long been bitter rivals in this region and the English East India Company wanted to make sure that the shipping routes between India and the Far East were secure. When Napoleon conquered Holland, the British occupied the Dutch possessions in the East Indies. Raffles played a leading role in these events. In the middle of organizing military operations and administrative systems in faraway islands, the extraordinary man found the time to observe the flora and fauna, record local customs and study ancient ruins.

After Napoleon was defeated, the Dutch wanted their colonies back. There were heated negotiations between Calcutta and Batavia (the Dutch headquarters, now Jakarta). The Dutch would eventually get back most of their possessions as per the Anglo-Dutch Treaty of 1824, but not before Stamford Raffles had made sure that the British would continue to control the Straits of Malacca.

The key to this strategy was the establishment of a new British outpost in Singapore. The island had been, in name, under the control of the Sultan of Johore but Raffles was able to secure it in exchange for the payment of an annual rent and British support against the Sultan's local rival. Singapore was formally founded on 6 February 1819 with a great deal of pomp and the firing of cannon.

Raffles is known today as the founder of Singapore but he had an extraordinary interest and curiosity about the natural and cultural history of South East Asia. He collected samples of plants and animals and even sent back a Sumatran tapir for the Governor General's garden in Barrackpore! He wrote about the Indianized culture of Java and Bali and is said to have 'rediscovered' the great stupa of Borobodur during the British occupation of Java.

Just before he returned to England, Raffles set up an institute in Singapore inspired by Calcutta's Fort William College. It survives as the Raffles institution, an elite school, though its original location on Bras Basah Road today is occupied by the Raffles City Shopping Mall, just across from the famous Raffles Hotel. There are so many places in Singapore today that are named after Raffles that it can be quite confusing for a visitor!

It's a Tiger! It's a Map!

As the British settled down firmly in India, they quickly discovered the need for good maps charting the country's interior in order to help with the administration, revenue collection and military movements. Till the mid-seventeenth century, European map-makers had been focusing on the coastline but now, the interiors also had to be mapped.

The tool used for doing this survey was the perambulator —a large wheel set up to allow the measurement of distance.

East India Company troops would often take a perambulator along on marches and estimate the distance by adjusting for the twists and turns of the road. This was not exactly accurate but it gave them readings that were a lot better than earlier estimates. For example, a map of Sri Lanka and the Coromandel coast from these times carries the note: 'The route from Tritchinapoly to Trinevelley ascertained by a march of English troops in 1775.' This was quite common!

The British decided to carry out a more scientific survey of Bengal after they had conquered it. In 1765, Robert Clive assigned James Rennel, a young naval officer, the task of making a general survey of Bengal. Rennell took a band of sepoys and travelled the countryside for seven years fixing latitudes, plotting productive lands and marking rivers and villages. It was hard and dangerous work. There were tigers everywhere and Rennell was only too aware of what they could do—he jotted down his fears in his notebook.

A tiger did carry off a soldier on at least one occasion. On another, a leopard jumped out of a tree and mauled five sepoys before Rennell grabbed a bayonet and thrust it into the beast's mouth! On yet another occasion, he was deeply wounded while fighting off bandits. At thiry-five, Rennell returned to England and produced the famous *Bengal Atlas*. He was hailed as the 'Father of Indian Geography'.

Though it was the best that had been done so far, Rennell had only covered a small part of the subcontinent. As British conquests expanded, the need for further surveys was felt. The task fell on William Lambton.

Lambton had had a long but ordinary career in India till he was made the Superintendent of the Great Trigonometrical Survey of India. This happened by chance. In 1798, he was sailing from Calcutta to Madras on a ship. There was a young colonel on the same ship called Arthur Wellesley who would later go on to become the Duke of Wellington, the victor at Waterloo. But in 1798, he was better known as the younger brother of the Governor-General and he was on his way to fight against Tipu Sultan of Mysore. Wellesley was impressed with Lambton and took him along for the expedition. Tipu Sultan was defeated and killed at the siege of Srirangapatnam. Lambton played an important role in this battle.

It was during this campaign that he came up with the idea of doing a survey of India using triangulation. This means, one takes three visible points as the corners of a triangle. The points should be visible from each position. If one knows the length of any of the sides and can measure the angles, the length of the other sides can be calculated using trigonometry. With the new measurements, another new triangle can be made and so on. This was tiring work but it provided very accurate measurements.

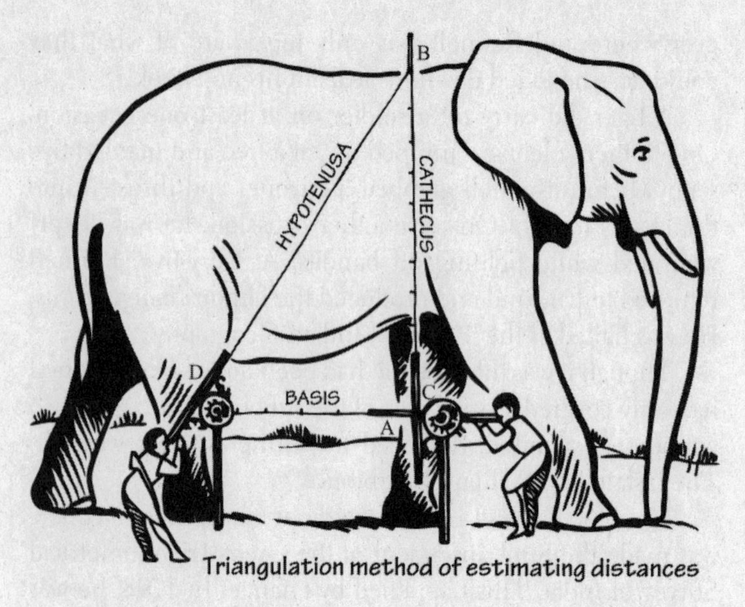

Triangulation method of estimating distances

Lambton followed this method to create an accurate map of India and also to use the measurements to establish the exact shape and curvature of the earth. This was not just out of scientific curiosity—it was of great importance to Britain, a naval and trading power. Lambton told Wellesley about his plan and Wellesley spoke to his brother, the Governor-General. And that's how Lambton landed his job!

The first thing that Lambton did was to order a modern theodolite to help with the survey. A theodolite is like a telescope that can help make very accurate measurements of angles needed for triangulation. The equipment Lambton ordered weighed half a ton and had to be shipped from England. On the way it was captured by the French and taken to Mauritius! But when the French realized that it was a scientific instrument, they very politely repacked and sent it to Madras. At last, Lambton could start on his work.

He began by establishing a baseline at sea level in 1802. He did this just south of Chennai's famous Marina Beach. From a flagpole on the beach, he found out the horizontal distance to the grandstand of the Madras racecourse. Once this was done, he started the sequence of triangulation that would criss-cross India for the next sixty years! This process lasted not just for his lifetime but for also that of George Everest, who took the job after him. In 1802, the East India Company had thought this work would be done in five years. The fact that this project was allowed to continue despite the time and resources it took up shows how important and useful it was considered by the British.

Carrying the heavy theodolite through jungles, mountains, farmlands and villages must have been very difficult. Often, there were bandits, local people who were hostile to the British, and kingdoms that had not made their peace with British rule. often, there were long delays because dust and haze made it difficult for them to take readings. At each location, the theodolite had to be dragged up to a height in order to provide a reading. Tall buildings were used when there were no hills. In 1808, Lambton decided to use the massive eleventh-century Brihadishwara temple in Thanjavur. This Shiva temple had been built by the Cholas at the height of their powers. It's a huge structure even by modern standards.

Unfortunately, the ropes slipped and the theodolite was smashed! Though it was so huge, it was a delicate instrument. Anyone else would have given up but not Lambton. He ordered a new one from England at his own expense and then spent the next six weeks repairing the damaged equipment with great pains.

Lambton worked on the survey till he died of tuberculosis in 1823. His forgotten grave was recently discovered by

writer John Keay in the village of Hinganghat, fifty miles south of Nagpur. His theodolite is in better condition and is now housed in headquarters of the Survey of India in Dehradun. Less than half of the project had been completed when Lambton died. But fortunately, George Everest was equally committed. By the time Everest retired and returned to England in 1843, the Great Arc had been extended well into the Himalayas.

Everest built a bungalow for himself at Hathipaon near Mussourie. The ruins of this bungalow still stand on a ridge with a magnificent view of snow-capped peaks on one side and the valley of Dehradun on the other. It is just a fifteen-minute drive from Mussourie town, followed by a ten-minute walk up the hill. Everest returned home a famous man and was knighted.

Did you know?

In 1849, the highest mountain in the world was discovered. It was more than 29,000 feet high! This mountain, Peak XV, was called Chomolungma or Mother Goddess of the World by the Tibetans. The Survey of India usually retained the local names for places wherever possible but not this time. The highest mountain in the world was renamed after George Everest. Yes, Peak XV is Mountain Everest.

The Revolt of 1857

By the time Mount Everest was named, the British were in control of the whole subcontinent. What was not directly ruled by them was managed through one-sided treaties with the local princes. Nobody else had controlled such a large part of the subcontinent since the Mauryans.

How did the British succeed in doing this when other Europeans had failed? It's true that they had the technology but this wasn't the only reason. It's not as if the technological gap between the Europeans and the Indians was as large as in the Americas or Africa. And there were vastly larger numbers of Indians than Europeans. There were also European armies for hire and allies fighting on the Indian side at times. And yet, the British were able to beat off much larger armies and still maintain control with a small number of officials. How?

What is surprising about the British conquest of India is that so few British were involved! The armies of the East India Company were mostly made up of Indian sepoys. In many cases, the British actually got support from the locals. For example, at the Battle of Plassey, Robert Clive was funded and encouraged by the merchants of Bengal. Some historians feel this shows, once again, that Indians did not think of themselves as a nation till the nineteenth century. But we've seen that this wasn't true and that Indians have had a strong sense of being a civilization for many, many years. Why did they not oppose the British rule more strongly then?

It's possible that this happened because the collapse of the Mughal empire in the eighteenth century had left the country in chaos. It had seemed that the Marathas would replace the Mughals, but they failed because of the loss at Panipat and internal fighting. The countryside was full of bandits and robbers. Some of them, like Begum Samroo, became so powerful and rich that they lived openly and in style in Delhi and were considered 'respectable' members of society.

The East India Company was not kind or generous but it did create some order in the country. Also, unlike the Portuguese, the British did not try and interfere with the local culture and social norms. Even when they did, like in the case of abolishing sati, they did it with the support of Indian reformists. This is probably why they didn't initially seem threatening to the Indians.

After his great victory at Plassey, Robert Clive did not offer thanksgiving at a church. He did it at a Durga Puja organized by Nabakrishna Deb in Kolkata!

But by the mid-nineteenth century, this open attitude changed. The British began to look at Indians as people who needed to be 'civilized'. They felt Hinduism was a 'superstition' and that the locals needed to be converted to Christianity if they were to be 'saved'. The Indians—both Hindus and Muslims—did not take well to this for obvious reasons.

This anger finally led to the Revolt of 1857, exactly a hundred years after the Battle of Plassey. Within a few weeks, the bulk of the East India Company's Bengal Army was in open revolt and, in many cases, the British officers had all been killed. This revolt spread like wildfire across large parts of north and central India. The revolt didn't have a single leader or a single group of leaders who were issuing orders. There were different centres with a number of different leaders, usually people from the old Indian aristocracy who had had their powers taken away from them.

Delhi was one such important centre of the uprising. By 1857, Shahjahanabad was no longer the glorious city it used to be. The eighty-two-year-old Bahadur Shah Zafar was an emperor only in name. The royal family survived on a pension the British gave them and many of the junior branches of the family were living in extreme poverty. William Sleeman, an official who visited the Red Fort a few years before the revolt, says that 1200 members of the family lived in the palace on the small pension but they were too proud to work! Instead, they would try and use their family name to cheat and make money. Even the palace inside the Red Fort was in shambles. In 1824, Bishop Herber described the palace gardens as 'dirty, lonely and wretched; the bath and fountain dry; the inlaid pavement hid with lumber and gardener's sweepings, and the walls stained with the dung of birds and bats'. Things would have been worse by the 1850s.

Writers like William Dalrymple have tried to present the court of Bahadur Shah Zafar as a 'court of great brilliance' and as a place that led to cultural growth but this is not exactly true. The court did have some excellent poets like Ghalib and Zauq but Delhi in those days was not Calcutta, where new ideas and innovations were afloat. Ghalib's poetry is beautiful but it's a lament about the world collapsing around him, not a vision of the future.

In May 1857, several hundred sepoys and cavalrymen rode into Delhi from Meerut and encouraged the local troops to join them. Together, they killed every British person they could find. Indians who had converted to Christianity were also killed. As more and more rebels arrived, the soldiers turned to the ageing emperor for leadership. Bahadur Shah was not sure what to do—if he listened to the soldiers, he was worried that the British would take revenge on him. But if he didn't, there was a large and growing number of angry men he'd have to face. He decided to play along with the rebels but he remained uncertain about his moves throughout this episode.

Meanwhile, a small British force arrived and set up a position of defence on the ancient Aravalli ridge overlooking the walled city. From here, they pounded Shahjahanabad with cannon. The British were few in number but the rebels were not well-organized and they could not capture the position from the British. A small group of Gurkha soldiers held off waves of rebel attacks near Burra Hindu Rao's house on top of the ridge—the place is now a hospital. One of the princes, Mirza Mughal, did try to organize the rebels but his efforts went in vain because the emperor couldn't make up his mind and members of his own family tried to pull him down. The British had a constant flow of information about what was happening inside the Red Fort throughout the siege!

In mid-August, the British received fresh troops and new supplies from Punjab. The attack became more intense. A month later, Shahjehanabad was captured and sacked. Bahadur Shah and members of his family, the proud descendants of Taimur the Lame and Ghengis Khan, fled down the Yamuna to take shelter in Humayun's grand tomb. They were soon discovered. Many Mughal princes were executed. Three of them, including the brave Mirza Mughal, were shot dead near the archway still called Khuni Darwaza or Gate of Blood. The emperor was exiled to Rangoon.

The city of Delhi was in ruins. Within the Red Fort, many of the Mughal structures were destroyed to make way for the barracks, which we can still see today. A few years later, a large part of the old city was cleared to build the railways. Only a few structures remained to remind one of Shah Jehan's dream. This was the end of India's third cycle of urbanization. It began with the sacking of Prithviraj Chauhan's Delhi and ended six and a half centuries later with the sacking of Mughal Delhi.

The next cycle, however, had already begun in Calcutta, Madras and Bombay. After the fall of Delhi, the British put down the other centres of rebellion one by one. Tens of thousands were executed as punishment. There were many extraordinarily brave people such as Rani Laxmibai of Jhansi who waged war against the British but the rebels were too uncoordinated to win.

Did you know?

Rani Laxmibai was only tweny-two years old when she fought the British, the most powerful empire of that time. For all its fame, the fort at Jhansi is a modest one. It still has two of Rani Laxmibai's cannon, of a design that was old-fashioned even in the mid-nineteenth century. It stood no chance against the British and yet, she had the audacity to defy them!

Many Indians were either indifferent to this rebellion or supported the British. Maybe they thought that if the British left, India would once again become chaotic like it had been in the eighteenth century. Maybe they thought the future did not lie in the old order of rulers. The year 1857 saw another kind of revolution. Three federal examining universities on the pattern of London University were established in the cities of Calcutta, Bombay and Madras. By the time India became independent in 1947, twenty-five such institutions were set up. The universities created an educated middle class that formed the forefront of the next round of resistance to British rule.

The rebellion of 1857 brought the East India Company to its end. Its territories in India were put directly under government control. The Governor-General was replaced with a Viceroy, a representative of the Crown. The ratio of Europeans to Indians in the army was pushed up to 1:3 from 1:9. That is, for every three Indians, there was now one European in the army.

The British stopped taking over Indian kingdoms and instead gave them a permanent standing under the Crown. This framework survived till 1947. The Queen's Proclamation of 1858 stated that the British would no longer try and impose their religion and customs on the local people.

Interestingly, the Queen's Proclamation was read out not in Calcuatta, Bombay, Madras or Delhi. It was read out in Allahabad, at the Triveni Sangam, the place where the Yamuna meets the Ganga and is said to be joined by the invisible Saraswati flowing underground. It is here that Ram is said to have crossed the river and visited the sage Bharadhwaj before going on exile to the forests of central India. There is even a tree under which Ram is said to have rested. It was also here that Xuan Zang saw the great gathering of the Kumbha Mela in the seventh century CE. Overlooking the temple and the merging rivers is the fort built by Emperor Akbar which has a Mauryan column with the inscriptions of three emperors—Ashoka, Samudragupta and Jehangir. In short, this was the heart of Indian civilization. The British seemed to have finally understood the nature of Indian nationhood.

If you visit the Saraswati Ghat in Allahabad at dawn in January during the annual Magh Mela or the Kumbha Mela, you can see tens of thousands of people of all ages, genders, classes, castes and sects take a dip in the confluence of rivers. They chant Vedic hymns composed thousands of years ago on the banks of the ancient Saraswati, still alive in the memory of the people.

The column built in memory of the Queen's Proclamation is a short walk from Saraswati Ghat and stands neglected in an overgrown park. The locals have forgotten about the significance of the place. This is sad because the modern Indian State is the direct outcome of this Proclamation. After Independence, the government capped the column with a replica of the national emblem, the Mauryan lions and the wheel.

Although colonial expansion became less open after 1858, a large gap remained between the Indians and the British. This is visible even in urban planning. British towns were spacious 'white-towns' while the towns of the locals were crowded 'black-towns'. It is not unusual for rulers to live separately from the ruled. We notice this in the citadel of Dholavira as well as the Red Fort of Shahjehanabad. But both sections still lived within the same cultural context. However, there was now a large cultural gap between the British and the locals.

It would be many decades before a small bunch of Indians with a Western education was allowed into places like the Civil Lines of Allahabad. Till recently, remains from this era were still visible in the large, crumbling bungalows of Allahabad's Civil Lines. But now, the area is turning into a jumble of malls, shops and apartment blocks.

The Steam Monsters

By 1820, India's population was 111 million but its share in world GDP had fallen to 16 per cent. China's share at this time was 33 per cent! Combined, they still accounted for half of the global economy. Because of the Industrial Revolution, Britain enjoyed a per capita income that was three times higher than that of the two Asian countries. This means that though the GDP shares of India and China were greater than that of Britain's, the people of Britain, on an average, were better off than the Asians. As the nineteenth century wore on, the gap between the Europeans and the Asians became wider. By the time India became independent in 1947, its share fell to a mere 4 per cent of the world GDP.

Though India's share had gone down, the second half of the nineteenth century saw big changes in the country's economic and geographic landscape. How did this happen? The British introduced the railways in India! There were many reasons behind this—some to do with trade and some to do with the military. Through the 1830s and 1840s, there were many discussions and proposals for the project. The government didn't have enough resources to take up something so huge and they thought they could ask private operators to raise the money. But not many were interested.

The discussions went on for several years and then came F.W. Simms, a railway engineer. A number of routes were surveyed under his supervision. He argued that a Delhi-Calcutta line would save the military at least 50,000 pounds a year, a very large sum in those days. The government decided to give generous guarantees to convince investors to put in money into the railways.

The very first railway line in the subcontinent ran 21 miles (34 km) from Bombay to Thane. The formal inauguration took place at Bori Bandar on 16 April 1854 when 14 carriages with 400 guests left the station 'amidst the loud applause of a vast multitude and the salute of 21 guns'. A year later, a train left Howrah (a town across the river from Calcutta) and steamed up to Hooghly. This was the first line in the east. Two years later, the first line in the south was established by the Madras Railway Company. By 1859, there was a line set up between Allahabad and Kanpur

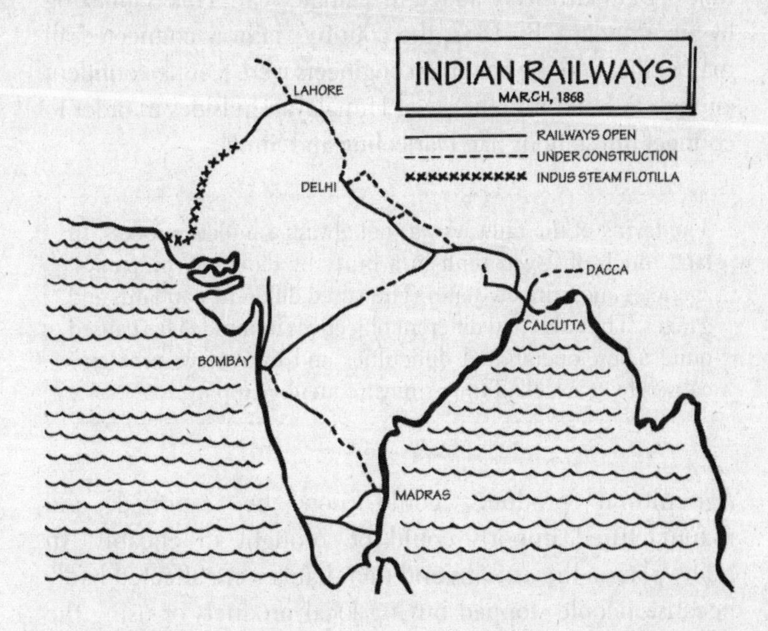

An Indian railways map of March 1868 shows that by this time, Howrah (i.e. Calcutta) had been connected to Delhi and this line then went on to Lahore. The Lahore-Multan line had also been built, some of it with the use of the four-thousand-year-old Harappan bricks you read about earlier. From Multan, you could use the Indus Steam Flotilla to sail down to Karachi. In the west, Bombay was connected to Ahmedabad and Nagpur but the link to the Delhi-Calcutta line was still not complete. The link between Madras and Bombay was also still being built near Sholapur. There were a number of side lines that were already being used or were being built.

All this work was happening with the technology available at that time and the difficult terrain of central India. In spite of this, work proceeded at a brisk pace and in the 1870s, an average of 468 miles (749 kms) was being added per year. In 1878, 900 miles (1440 km) were added in a single year! This is amazing by any standard. By 1882, the country's railway connected all major cities and the Victorian engineers were feeling confident enough to build into the steep Himalayan hillsides in order to connect hill stations like Darjeeling and Simla.

The laying of the railways was not always a smooth process. In fact, much of it was built in a hurry by different companies, agencies and princely states. They used different standards and gauges. They also had different objectives in mind. This caused quite a few operational difficulties and even now, problems caused by this lack of uniformity frequently crop up.

Agricultural products could now be exported and manufactured imports could be brought in cheaply. In many places, the artisans and merchants were affected badly because people stopped buying local products or using the old caravan routes. The Marwari merchants of Rajasthan, for example, were forced to leave their homes and look for opportunities elsewhere. Many moved to Calcutta where their descendants became successful businessmen. Their old ancestral homes can still be seen in towns like Mandawa and Jhunjhunu in the Shekhawati region of Rajasthan.

In the meantime, new towns came up along the railway routes and some communities took advantage of this rapid growth. One such group was the Anglo-Indians or the Eurasians who joined the railways in large numbers. At one point, the Anglo-Indians of India had their own distinct culture, with their own cuisine, love for music and sport, and their way of speaking English. But now, they have more or less merged into the Indian Christian population. Many of them migrated to Australia and Canada and they, too, have become one with the population there.

Just as the Internet or mobile technology now connects us all, the railways connected people in those times. Since it carried people and goods from across the country, it allowed them to come together and interact.

The social reformer and religious leader Swami Vivekananda used trains to travel the country in the last decade of the nineteenth century. Mahatma Gandhi did the same as he tried to get to know India again after he returned from South Africa.

By 1924, 576 million passenger trips were being made per year. Of course, this does not mean that train journeys were always enjoyable. Especially for the second and third class passengers. A report listed out the following complaints of third class passengers in 1903:

- Overcrowding of carriages and insufficiency of trains
- Use of cattle trucks and goods wagons for pilgrims
- Absence of latrines in the coaches
- Absence of arrangements for meals and insufficient drinking water
- Absence of comfortable waiting halls
- Inadequate booking facilities
- Harassment at checking and examination of tickets
- Bribery and exactions at stations, platforms and in the train

- Want of courtesy and sympathetic treatment of passengers by railway staff

Sound familiar? Many of these complaints still ring true! Thankfully, people are not made to travel in cattle trucks now but the phrase can still make one angry—remember Shashi Tharoor made a comment about travelling in 'cattle class' when he was minister, and got into trouble!

Bombay Then and Now—Still the Same!

The period between the Revolt of 1857 and the First World War was the time when the British were at their most powerful. This was obvious in Calcutta, the empire's eastern capital, where luxurious and large buildings were built by the government, banks, companies and the rich people. Many of these, like the High Court, the Writers' Building, the Chartered Bank Building, the General Post Office and Guillander House, still exist. There's also the Raj Bhavan, once the palace of the Governor-General, which is now home to the Governor of West Bengal.

Just as Calcutta was basking in all this attention, Bombay, too, was becoming a very important city for the British. Bombay was not a new settlement. The area had been a major port even in ancient times; the seventh-century cave temples of Elephanta Island tell us this. But the modern city originated from when the Portuguese occupied the area in the sixteenth century.

At this stage, Bombay was a group of several marshy islands. The name of these islands are still used in Bombay as names of neighbourhoods—Colaba, Mahim, Parel, Worli, Mazagaon. The islands passed into British hands in 1662 as part of the dowry received by King Charles II on his marriage to Catherine of Braganza. They were then leased to the East India Company for ten pounds a year.

In the beginning, the Marathas prevented the British from expanding into the mainland. But by the late eighteenth century, the British were in a strong position and Bombay became an important port for trading. The British governor started a series of civil engineering works, loosely called the Hornby Vellard Project, to connect the various islands by landfills and causeways. By 1838, the seven southern islands were combined to form a single Bombay Island. By 1845, the Mahim causeway connected Mahim to Bandra on the island of Salsette. Though the main islands have all been joined, linkages are still being built to this day. The latest is the Bandra-Worli Sealink, which opened in 2009 to connect South Mumbai to the suburbs.

> One of the first people to take advantage of the new Bombay were the Parsis, descendants of Zoroastrian refugees from Iran who had settled along the Gujarat coast. They first moved to Bombay to work for the British as shipbuilders but, by the 1830s, grew very rich by becoming part of the opium trade with China.

In the mid-nineteenth century, Bombay was smaller than Calcutta or Madras. But in the 1860s, things changed. The American Civil War broke out and the American North blocked the ports of the American South. This meant that the mills of Lancashire, in England, could not get raw cotton. And so, they looked to western India for the material. The newly built railway network transported cotton directly from the fields to the Bombay port. New cotton mills began to be built in Bombay itself. The opium trade with China also boomed at the same time, with 37,000 chests being shipped out every year.

With all this new money, both the government and the wealthy merchants of the city started building new structures; the grander the better. The cotton trade was booming. The buying and selling of land had also become a profitable business. An informal stock exchange appeared under a tree in front of the Town Hall. People from other parts of the country moved in by tens of thousands, and crowded slums came up.

A traveller writing about this commented, 'To ride home to Malabar Hill along the sands of Back Bay was to encounter sights and odours too horrible to describe… to travel by train from Bori Bunder to Byculla, or to go to Mody Bay, was to see in the foreshore the latrine of the whole population of Native Town.' The location of the slums of Bombay have changed over the last one-and-a-half centuries but anyone who has travelled in Mumbai's suburban trains will know what the above comment means!

In 1865, the American Civil War ended and the prices of shares and cotton in Bombay crashed. By 1866, several of the city's banks and real estate companies failed and many rich people were left without any money. The city was full of half-built projects that were abandoned. But the boom years had given Bombay a new status and even now, the spirit of those times is alive. Strike up a conversation with the street vendors of Nariman Point or the Fort and they will give you tips for the stock market!

The Man Who Floated Logs

By the 1860s, the British surveyors had an accurate map of the subcontinent and were beginning to wonder what was there beyond the Himalayas. This was not just because they were curious; it was because the Russians had started to invade Central Asia. The 'Great Game' had begun.

The problem was that the Tibetan authorities did not want to let in Europeans inside their borders—a few who had tried had been tortured and killed. The Survey of India decided to use Indian spies disguised as traders and pilgrims. The first among this group was a young schoolteacher from the Kumaon hills, Nain Singh. In 1865, he crossed from Nepal into Tibet along with a party of traders. A few days after the crossing, the traders slipped away one night with most of Nain Singh's money, leaving him alone in a strange land.

Fortunately, they hadn't stolen his most valuable possessions, concealed in a box with a false bottom—a sextant, a thermometer, a chronometer, a compass and a container of mercury. He also had with him a Buddhist rosary but this one had 100 beads instead of the usual 108. Nain Singh planned to measure distance by slipping one bead for every 100 paces walked. He also had a prayer wheel which contained hidden slips of paper on which he recorded compass bearings and distances.

Nain Singh begged his way across the cold and empty landscape. In January 1865, he finally entered the forbidden city of Lhasa. He pretended to be a pilgrim and even made a brief visit to the Dalai Lama of that time. He supported himself by teaching local merchants the Indian system of keeping accounts. But he knew that his life would be in danger the minute he was discovered—after all, he'd witnessed the beheading of a Chinese man who had arrived in Lhasa without permission! Nain Singh stopped appearing in public too often after this incident. At night, he would climb out quietly from the window on to the roof of the small inn where he stayed. Then, he would use his sextant to determine latitude by measuring the angular altitude of the stars. He also used his thermometer to record the boiling point of water as the higher the altitude, the lower the boiling point. Using this method, he estimated that Lhasa was about 3420 metres above sea level. The modern measurement is 3540 metres—not bad, huh?

Nain Singh left Lhasa in April along with a Ladakhi caravan and headed west for 800 km along the River Tsangpo. All along, he kept taking readings in secret. After two months, he slipped away on his own and made his way back to India through the sacred Mansarovar Lake. He came back to the Survey of India headquarters on 27 October 1866. During his twenty-one-month adventure, he had surveyed thousands of kilometers, taken thirty-one latitude fixes and determined height in thirty-one places. And he'd determined the first accurate position of the Tibetan capital! Nain Singh later returned to Tibet and explored a more northerly route from Leh in Ladakh to Lhasa. Some members of his family joined the dangerous profession and went on to work for the Survey of India.

Nain Singh's reports raised an important geographical question. Where did the Tsangpo flow? Did it cross the Himalayas as Singh suggested? Was it the river known to Indians as the Brahmaputra? To solve this mystery, the people at the Survey of India decided to send someone back into Tibet and float something identifiable down the Tsangpo. If it turned up in the Brahmaputra in Assam, they'd know the answer!

The two-man team for the job was made up of a Chinese lama living in Darjeeling and a Sikkimese surveyor called Kinthup. But the lama was quite the man of jolly times. He was more interested in getting drunk than doing any serious work. The team was stuck in one village for four months because the lama fell in love with their host's wife! When this story came to be known, he had to pay Rs 25 in compensation and leave the place.

Things did not improve when at last the team crossed into Tibet. The lama sold Kinthup as a slave to the headman of a Tibetan village and disappeared! From May 1881 to March 1882, Kinthup worked as a slave before running away to a monastery. After living for several months as a monk, he received permission to go on a pilgrimage. He went to a place near the Tsangpo and spent many days cutting up 500 logs into a regular size. He hid these in a cave and then returned to the monastery.

A few months later, he received permission to go to Lhasa on a pilgrimage. There, he received a fellow Sikkimese to write a message to his bosses at the Survey. He told them what the lama had done to him and then said that he had prepared 500 logs according to the orders given to him. He was going to throw 50 logs a day into the Tsangpo from Bipung in Pemake, from the fifth to the fifteenth day of the tenth Tibetan month of the year called Chuhuluk, of the Tibetan calendar.

Kinthup did what he'd promised to do. But the watch on the Brahmaputra had been abandoned and the letter came too late. We do know now that the Tsangpo is indeed the Brahmaputra. The logs must have floated down to Assam and then Bengal. Kinthup did not become famous as he deserved to. He spent his remaining life as a tailor in Darjeeling. These were the days of adventure that writers like Rudyard Kipling captured in books like *Kim* and *The Man who would be King*.

The Last of the Lions

The British didn't just take surveys and build large structures in India. They also had a good time! One of their popular pastimes, like the rulers before them, was hunting, especially tiger hunting. According to Valmik Thapar, as many as 20,000 tigers were shot for sport between 1860 and 1960 by Indian princes and British hunting parties. Another estimate says that about 80,000 tigers may have been killed between 1875 and 1925. Tigers were thought to be dangerous animals and rewards were given to those who killed them. Despite this mass killing, the tiger population in 1900 was between 25,000 to 40,000. But where were the lions?

You may remember Sir Thomas Roe who had to obtain special permission from Emperor Jehangir to hunt a lion that was troubling his group. Clearly, the British did know about the animal. There are many accounts from Aurangzeb's time which suggest that the lion was still quite common in the beginning of the eighteenth century. But their numbers seem to have suddenly fallen by the mid-nineteenth century. Why did this happen?

First, it's possible that modern guns led to the lion's downfall. It became easy to kill an animal that lives in the open. Second, with the fall of the Mughals, lion hunting was no longer restricted to the royals. Anybody with a gun could go out and hunt the animal. Despite this, there were still a number of lions in North India in the early 1800s. William Frazer is supposed to have shot eighty-four lions in the 1820s and he took great pride in having been personally responsible for the extinction of the species in Haryana. There are reports of large lion populations in central India in the 1850s and of ten lions being shot in Kotah, Rajasthan in 1866. Then, suddenly, the lions simply disappeared except for a small population in Gujarat. What happened?

Could it be that habitat loss had led to their disappearance? According to Angus Maddison's estimates, between 1820 and 1913, India's population jumped from 209 million to 303 million (not counting the rest of the subcontinent). To feed this huge population, it became necessary to increase farming. The railways made it possible to export agricultural products like opium and raw cotton. So the open ranges needed by the lion (and the cheetah) were just gobbled up by farming in a few generations. The tiger, too, lost much of its habitat but it could live in hilly and swampy terrain and so it survived better.

By the late nineteenth century, there were reports that perhaps only a dozen Asiatic lions were left in the wild in the Gir forests of Junagarh, a princely state in Gujarat. The actual number was probably higher but finally, people began to worry about the lion. Lord Curzon, the Viceroy, refused to go on a lion hunt in Gir during his state visit to Junagarh in November 1900. The Nawabs of Junagarh, with the support of the British government, now became the guardians of the lion for the next half-century.

The Gir forest was protected and hunting was strictly regulated. Only the most senior British officials and Indian princes were allowed to hunt them. The Nawabs took their job very seriously—they refused permission to many princes and British officials though it wouldn't have been easy for them to do so. Gir is still the only place where the Asiatic lion survives in the wild—a count of 411 in 2010.

The Indian cheetah was not as lucky as the Asiatic lion. The last documented sighting of the animal was in Madhya Pradesh in 1947, the same year that India became independent.

A New New Delhi

After the sack of 1858, Delhi became a mere district headquarters in Punjab province. The 1881 census shows that its urban population had come down to 173,393. The Mughal-era city of Shahjehanabad was still the main urban hub, with European troops based inside the Red Fort and Indian troops in Daryagunj. The railways connected the city to Lahore in the west and to Calcutta in the east. To the north of the walled city, the British had built a Civil Lines with large bungalows and gardens. With all its old ruins, Delhi in the late nineteenth century would have been beautiful but really, compared to Bombay, Calcutta or Madras, it was not a 'happening' place. And so it remained till 1911.

In the meantime, tiny cracks were appearing in the British Raj. Yet again, nature played a role in this. From 1874, India suffered a series of severe droughts. At first Bengal and Bihar were affected but Viceroy Lord Northbrook and Famine Commissioner Sir Richard Temple dealt with it by importing rice from Burma. But the British government, headed by Prime Minister Disraeli, didn't approve of this. They said that Northbrook and Temple had wasted money! Northbrook resigned over this. Lord Lytton took over from him.

In 1876, the rains failed for the third time and the famine situation became really bad in southern India. Lord Lytton, however, was not moved to action. He even scolded the Governor of Madras for being too generous. Sir Richard Temple, in the meantime, had learnt his lesson. He did not intervene as he had earlier. By 1877, the famine spread across the Deccan and Rajasthan to the north-west and yet grain from places which had surpluses was still being exported out to the rest of the world.

The Great Famine directly or indirectly killed 5.5 million people, more than two-thirds of them in British-controlled parts of the subcontinent.

In the middle of such a terrible crisis, when people were dying all around him, Lord Lytton blew up a lot of money on the Delhi Durbar of 1877 where Queen Victoria was proclaimed the Empress of India in front of all the princes of the subcontinent. This show of arrogance led to a lot of anger which ultimately led to the formation of the Indian National Congress in 1885.

As the demands for independence became louder, the British government decided to take steps to prove that they had a right to rule over the country. One idea was to follow the Mughals and build a new capital in Delhi because it was argued that the 'idea of Delhi clings to the Mohammedan mind'. Viceroy Hardinge thought this was his best chance to be remembered as the founder of a great city. The British also needed a grand sound bite for the Durbar held in 1911 to commemorate the coronation of George V as Emperor of India. The proclamation was read out at Coronation Park, to the far north of the city. This is the same spot where Queen Victoria had been declared the Empress of India. A great stone column was raised to mark the event.

Almost no tourist visits the place these days. King George V glares down from a pedestal removed from the canopy opposite India Gate in the 1960s. There are also several pedestals without statues, as if their occupants were upset that nobody was visiting them and simply walked off!

The architects Edwin Lutyens and Herbert Baker were given the job of designing the new city of Delhi. The original idea was to build the city to the north of Shahjehanabad, roughly around where Delhi University now stands. But after a number of ground surveys, it was decided that the new city would be built to the south of the existing urban centre. This area was close to the ruins of many older Delhis—Dinpanah, Indraprastha, Feroze Shah Kotla. The new city was not built for trade or industry. It was constructed to show the power of the British Empire.

The centrepiece was the palace of the Viceroy built on Raisina Hill—what we now know as Rashtrapati Bhavan. There were many opinions about what this building should look like. Should it be classical European? Indo-Sarcenic? Mughal? Lutyens's own opinion of Indian aesthetics was closer to those of Mughal Emperor Babur but Baker preferred the style of the locals. Ultimately, they decided on a design that combines classical European columns with Mughal and Rajput detailing. In front of the palace was a grand venue called Kingsway (now Rajpath) inspired by the Mall in Washington DC. The intention was to impress and more than a century later, it still impresses.

RASHTRAPATI BHAVAN

The rest of New Delhi consisted of government offices and big bungalows built like a garden city. It was a Civil Lines on a huge scale with a strict order of things. There was no space for senior Indian officials because the British never thought there would be one! The whole thing was designed for a population of 60,000 or lesser, including servants and other support staff. The only space for commerce was Connaught Place and its surroundings. Called 'Lutyens's Delhi', this city is today the capital of the Republic of India.

Did you know?

In those times, the senior white officers who were to live within the Civil Lines were informally called 'fat white'. It's funny that, after Independence, over-fed politicians who pretend to be poor in their white kurta-pyjamas should live in these spacious bungalows which were meant for the 'fat white' of those times!

A lot has been written about the grand buildings and bungalows of Lutyens's Delhi. But if you look at early photographs of the cityscape in the 1920s and 1930s, it looks very different from what we see today. It's not just that much of the city is under construction. You also see that the trees we now associate with the city have not yet grown! The systematic and careful planting of trees was a very important part of the overall design—it's a feature we still identify with Delhi.

The planting of trees was not new in Delhi. At its heights, Shahjehanabad (Old Delhi) had many private Mughal gardens belonging to the royal family and senior nobility. This included the Begum Jehanara's gardens north of

Chandni Chowk and the two famous gardens within the Red Fort—Hayat Baksh (Life-Giver) and Mahtab Bagh (Moonlit Garden). The British, however, took this to a totally different level as they tried to create a garden city. There were heated debates between foresters, horticulturists and civil servants about which species should be planted. Finally, the Town Planning Committee submitted a report in 1913 with a list of thirteen trees including neem, jamun and imli that were thought to be suitable for planting along the avenues of New Delhi. Other species were planted later but trees from this original list still dominate many of the roads of Lutyens's Delhi.

The British also spent a lot of money and resources on reforesting the Aravalli ridges around New Delhi, particularly the Central Ridge just behind Rashtrapathi Bhavan. The mesquite, a Central American tree, called the 'vilayati keekar' was the tree that was commonly planted. Many think this is a local tree but it's actually not and it has successfully pushed out many trees that are actually native to the place! As a result of all this tree planting, central Delhi looks really green when seen from a height.

As the construction of the new city drew close to completion, the British raised their own pillar in front of the Viceregal palace—the Jaipur column headed by a six-point crystal star. It is easily visible through the main gate on Raisina Hilla. At its base is the inscription: 'In thought faith/ In word wisdom/ In deed courage/ In life service/ So may India be great.' Was this a patronizing blessing or an expression of awareness that British rule would one day end? Did the British also leave behind a column like Ashoka so that future generations would think well of them? By the time New Delhi was completed in the mid-thirties, it was quite clear that British rule wouldn't last for long.

Ship Ahoy! Again!

As we have seen, India had withdrawn into itself from the twelfth century. Why did the caste rules that prohibited a person from crossing the seas come up? We don't really know. It's quite puzzling because many Indian merchants and princes became very wealthy because of overseas trade. Brahmin scholars also benefitted because there was a great demand for them in South East Asia.

Despite these rules, there were Indian Muslims and even Hindus who continued to travel to foreign lands. There are remains of a large Indian trading post in faraway Azerbaijan. Built in the seventeenth and eighteenth century, the Ateshgah of Baku includes the remains of a Hindu temple and inscriptions invoking the gods Ganesh and Shiva. There are also records of Indian merchants in Samarkand and Bukhara. Still, these outposts were not as busy as the ones that had existed in the past.

It was in the nineteenth century, under British rule, that Indians began travelling abroad again in large numbers. After the abolition of slavery in 1834, the British needed bonded labour in their colonies. In the beginning, this demand came from sugarcane plantations but soon, Indians were used to build railway lines and work mines. In the early years, the workers thought that they could return home after the period of their bond was over, but the British decided it would be cheaper if these people settled down in the colonies. Indian women were encouraged to join their husbands in these colonies. The bonded workers lived a hard life but more and more of them signed up for it because of the Great Famine of 1877.

Large Indian communities settled down in faraway British colonies like Fiji, Trinidad, Guyana, Malaya, South Africa and Mauritius. The French colony of Reunion and the Dutch colony of Surinam also had several Indians. The place where half a million Indian workers landed in Mauritius is called 'Aapravasi Ghat' or Immigration Depot and is now a UNESCO World Heritage Site.

Of the hundreds of thousands of Indians who left their homes with contracts, less than a third returned. Many died during the sea journeys and the years of hard labour. Yet, enough of them survived to form the Indian communities scattered across these faraway lands.

Soon, Indian traders and clerks also began to follow the British to the colonies. Gujarati merchants and shopkeepers established a network in eastern and southern Africa. The Tamil Chettiar community was especially active in South East Asia and established a network in Burma, Malaya, Singapore and even French-controlled Vietnam. As they settled in these places, they found tiny remnants of Indian merchant communities that had survived from ancient times! The Chitty community in Malaysia is one such example.

Though this network of Indian communities was created and controlled by the British, these communities played an important role in India's struggle for independence. Mahatma Gandhi, for example, was part of the Indian community in South Africa between 1893 and 1914. He developed his political and spiritual philosophy of non-violence while fighting for the rights of Indians there. The incident in June 1893 when he was pushed off the first-class compartment of his train despite having a valid ticket changed his life. This incident took place at Pietmaritzburg station—there is a plaque here today which shows where he was thrown out. Gandhi returned to India only in 1915, at the age of forty-six, but he soon became the country's leading political figure.

Singapore, by contrast, was the centre of a very different effort to free India. When the Japanese captured the city during the Second World War, Netaji Subhash Bose used the opportunity to form the Azad Hind Fauj or Indian National Army. This army was made up of Indian civilians and soldiers who were prisoners of war. The first review of the troops took place in July 1943 on the Padang, a large open field that still exists at the heart of the city. There is a small memorial near the Singapore Cricket Club that marks the event. The original one was demolished by the British after the war, so the current memorial dates only from 1995.

Near the memorial is Dhoby Ghat, where Bose declared the formation of the Provisional Government of Free India. This proclamation was read out at the Cathay Cinema Hall. The building has been demolished but a part of its façade exists as part of a new shopping mall.

Bose's army failed in military terms—the Japanese who sponsored it lost the war. But the effort succeeded in creating doubts among the British about the loyalties of their Indian troops. Seven decades have passed but there are still a few Singaporean and Malaysian Indians who were part of these events who are alive today. Isn't it remarkable that these people, many of whom had been born outside India and had never been there, were willing to die for the idea of India?

8
We're Munni and Modern

After so many years of foreign rule, India finally became an independent country on 15 August 1947. But with Independence came the Partition. The subcontinent was divided into Muslim-dominated Pakistan and Hindu-majority India. There was widespread violence throughout the country.

In addition, over a third of the country was ruled by local princes who did not necessarily want to hand over their kingdoms. Parts of it were still ruled by the French and Portuguese. The long border with China (which was originally a border with Tibet) was disputed. Thus, the borders of modern India were not clearly drawn in August 1947—it came to its current shape only in the mid-1970s when Sikkim was included in the Union. India still has border disputes with China and Pakistan, so even the borders we have now may change in future.

The Partition

The Partition of India happened because of the differences among political leaders about what makes up nationhood.

These were the same differences that Akbar and Aurangzeb had in their approach to the empire they were ruling. While one was tolerant of other people's faiths, the other believed in the superiority of one religion over others. The sixteenth-century Islamic scholar Ahmad al-Sirhindi, from Punjab, was very critical of Akbar's liberal attitude. This was the same kind of thinking that made Mohammad Ali Jinnah and the Muslim League demand a separate country for Muslims.

It was decided that India would be divided along religious lines. The meeting that finalized the Partition was held on 2 June 1947 in Viceroy Mountbatten's study, under a large painting of Robert Clive. The Indian National Congress was represented by Jawaharlal Nehru, Sardar Patel and Acharya Kripalani. The Muslim League was represented by Jinnah, Liaquat Ali Khan and Rab Nishtar. Baldev Singh represented the Sikhs. Lord Ismay and Sir Eric Mieville, two of the Viceroy's key advisers, were also present.

The decision to partition India was announced at 7 p.m. on 3 June on All India Radio. The Viceroy spoke first, followed by Nehru and then by Jinnah. Pakistan was about to be born. The date of birth, however, was not announced. But when Viceroy Mountbatten was later asked about it at a press conference, he replied that the final transfer of power to India would happen on 15 August—just seventy-two days later. It looks like Mountbatten took this decision himself—he had not asked the Indian National Congress, the Muslim League or even Downing Street about it. It came as a shock to everyone.

It is unclear why Mountbatten chose August 15 as the day of Independence. Could it be that he had a sentimental attachment to it? It was the same day that the Japanese had surrendered to the Allies in the Second World War just two years earlier.

The Partition of India was a major project. It was called the 'the most complex divorce in history' and the time to complete this divorce was really short! Everything from the apparatus of the State, including the army, to government assets and debts had to be divided fairly between the two new countries. Even chairs, tables, petty cash, books and postage stamps had to be divided. There were many arguments over the silliest of things. Sets of *Encyclopedia Britannica* in government libraries were partitioned. The instruments of the police band in Lahore were also divided up—a drum for India, a trumpet for Pakistan and so on! In the end, the last trombone was left and the two sides almost came to blows over it! Manto's short story 'Toba Tek Singh', which is about how the inmates of Lahore's mental asylum had to be divided up along communal grounds, captures the madness of these times accurately.

But all these petty fights were nothing compared to the real business of partitioning the land, particularly the two large provinces of Punjab and Bengal. It was Sir Clyde Radcliffe who had to do this. He was considered to be one of the most brilliant lawyers of his time but he had had nothing to do with India. And since he knew nothing about India, it was thought that he'd be the best person for the job—he wouldn't be partial! On 27 June, he was called to the office of the Lord Chancellor and assigned the task. Radcliffe must have been stunned when he heard this. He was being asked to decide the fate of millions of people with no previous knowledge of the land he was expected to divide. He must have known that whatever he did, nobody was going to be happy and there would be violence. It was the worst job in the world.

Radcliffe began work in the terrible July heat based in a lonely bungalow in the Viceregal estate in Delhi. There was barely any time for the big day, so he could not visit the lands that he had to divide. Instead, he had to trace out a boundary line on a Royal Engineers map with just population statistics and maps for company. This was really difficult. The Hindu and Muslim areas were mixed up randomly. The city of Lahore, for example, was split exactly between the Muslim and Hindu-Sikh populations—6,00,000 each. Amritsar was a holy city for the Sikhs but it was surrounded by Muslim-majority areas. In Bengal, Calcutta was the main industrial centre and had a Hindu majority. But the raw jute for its jute mills came from the Muslim-majority east! In Punjab, partition would mean that irrigation systems had to be broken up. How was he to manage these difficulties?

Even as Radcliffe was drawing his line, communal violence was escalating across the country. People had already started to move from one region to another even before the border had been drawn. Radcliffe's maps were sent to the Viceroy on 13 August, but they were not made public for seventy-two hours. So when India become independent on 15 August, many Indians along the borderlands did not know which country was now their home. The maps were made public a day later. And the bloodbath began.

People were on the move—on trains, on bullock carts and on foot—holding on to whatever they could save. About seven million Muslims moved from India to Pakistan and a similar Hindu-Sikh population moved from Pakistan to India. Radcliffe, who was tired and fed up of the process, returned to London. He returned the 2000 pounds that he had received for his services.

The Hindus and Sikhs who fled to West Pakistan were sent to hundreds of refugee camps. One of the largest camps was in Kurukshetra, the battlefield where the Pandavas and the Kauravas are said to have fought each other in the Mahabharata. The camp was planned for 1,00,000 but three times the number came to live in it by December 1947. Half a million refugees, mostly from West Punjab, arrived in Delhi. These desperate people squatted wherever they could, including the pavements of Connaught Circus. In time, they built homes in colonies given to them in the south and west of Lutyens's garden city. We know them today as Lajpat Nagar, Rajendra Nagar, Punjabi Bagh and so on.

A smaller group of refugees from East Pakistan also came to Delhi and were settled in the East Pakistan Displaced Persons Colony. Now named Chittaranjan Park, it still has a very Bengali identity. Within a few decades, Delhi went from being the city of Mughal memories to a grand colonial dream and then a city of refugees.

The migration happened in one big rush in Punjab but it happened over many years in Bengal. A series of anti-Hindu riots in East Pakistan in 1949–50 once again made people move from their homes. About 1.7 million came to West Bengal in 1950 alone. This migration continued for over a decade. People squatted wherever they could—in railway stations, unoccupied homes, vacant land and even barracks. There was anger that the national government in Delhi did less to help the Bengali refugees than they did the Punjabi ones. A large population of Hindus continued to live in East Pakistan. They would face another crisis two decades later.

Despite all their troubles, the Punjabis and Bengalis at least had provinces, West Bengal and East Punjab, to call their own. But there were many others who didn't know where they belonged. Like the Sindhis who now found that their entire province was under Pakistan. Many of them went to Bombay to live in refugee camps. Many Sindhis remain in Ulhasnagar, an industrial suburb of Mumbai. Over the years, they have migrated all over the world and today run a network of international businesses. Hong Kong, for example, has a number of successful Sindhi business families.

Princes in Distress

What about the provinces that were ruled by princes? There were over 500 of them. Some of the princes ruled kingdoms that were as big as major European countries while others ruled only a few villages. Some of these kingdoms had survived from before the time of the Mughals. It says a lot about the spirit of the times and the skills of the negotiators that despite all the grumbling and last minute bargaining, almost everyone signed over their kingdoms to the new democracy by the 15 August deadline. Some opted for Pakistan.

However, there were three important kingdoms which were not handed over—Junagarh in the west, Hyderabad in the south, and Jammu and Kashmir in the extreme north. The first two had Muslim rulers but a Hindu-majority population. Jammu and Kashmir had a Hindu ruler but a Muslim-majority population.

Junagarh was not just wedged within Indian territory, but it was also of great symbolic value. The ancient temple of Somnath and the sacred hill of Girnar with its many Hindu-Jain temples were here. At the base of the hill, and a short walk from the Junagarh fort, are the rock inscriptions of Ashoka, Rudradaman and Skandagupta that you read about earlier. It's also the last home of the Asiatic lions left in the wild. In 1947, it was ruled by Nawab Mohabat Khan, best remembered for his love of dogs. He apparently owned 2000 pedigree dogs and when two of his favourites mated, the 'wedding' was celebrated as a State event!

In the summer of 1947, Mohabat Khan was on holiday in Europe but he had left his kingdom in the hands of his

Dewan, Sir Shah Nawaz Bhutto, a Sindhi politician and the father of future Pakistani Prime Minister Zulfikar Ali Bhutto. When the Nawab returned, Bhutto convinced him to go with Pakistan. On 14 August 1947, just hours before the handover, Junagarh announced its decision. Pakistan accepted the decision a few weeks later.

The local population (82 per cent Hindu) as well as the Indian leaders were angry. Deputy Prime Minister Patel, a Gujarati, responded by getting two of Junagarh's vassal states to announce that they were part of India. A small military force was sent in to support them. In the meantime, people began to agitate and protest. The Nawab panicked and fled to Karachi, taking with him a dozen of his favourite dogs! Sir Shah Nawaz, who had no options left, agreed to an election in which the people voted in favour of India.

This chaotic period in Junagarh was bad news for the lions of Gir. With the Nawab's protection crumbling, several lions were hunted down in the later months of 1947. Some of the hunters were princes of neighbouring kingdoms who simply took advantage of the situation to add to their private collections. Thankfully, order was restored by early 1948. This was not just to conserve wildlife. The lion, as shown on the Mauryan pillar in Sarnath, was now the national emblem. There had been some who had wanted the elephant but a committee headed by future president Rajendra Prasad decided on the lion in July 1947. The same committee also decreed that the flag of the Indian National Congress would become the national flag after changing the symbol of the charkha (spinning wheel) to that of the spoked wheel from the same Mauryan column—the ancient symbol of the Chakravartin or Universal Monarch. After thousands of years, Sudasa's dream was still alive.

Meanwhile, a fresh new problem was brewing in Hyderabad, a leftover from Aurangzeb's invasion of southern

India. It was the largest of the princely states and its ruler, Nizam Osman Ali Khan, was famous as one of the richest and most miserly men in the world. Though the state was inhabited by a Hindu majority, the police, civil service and landowning nobility were full of Muslims. It even had a large army that included armoured units as well as Arab and Afghan soldiers for hire. When it became clear that the British were leaving, the Nizam first tried to make his kingdom an independent one and then hinted that he would opt for Pakistan.

This was a strange threat to make because Hyderabad was surrounded by Indian territory. Still, the Nizam was persuaded by Kasim Razvi, an Islamic fanatic, to allow the creation of an army called the Razakdars. There were about 2,00,000 of them at its height! As the political situation became worse, the Razakdars terrorized the countryside. India responded by squeezing Hyderabad out economically. Finally, in September 1948, more than a year after Independence, Deputy Prime Minister Patel decided enough was enough.

The military action to make Hyderabad a part of India was called Operation Polo, supposedly because of the large number of polo grounds in Hyderabad!

The Indian army entered Hyderabad on 13 September. The Razakdars as well as the regular troops of the kingdom put up a fight but everyone knew how it was all going to end. By the morning of 17 September, the results were clear. The surrender was surprisingly meek. *TIME* magazine reported that the commander-in-chief of Hyderabad's army, a black-moustached Arab called Major General Syed Ahmed El Erdoos, drove up in a shiny Buick to a place a few miles outside the city. He then walked up to Major General Chaudhuri, the Indian field commander. They apparently 'shook hands, lit cigarettes and talked quietly while the spellbound villagers looked on'.

But the story of Jammu and Kashmir is very different. Here, a Hindu prince ruled over a Muslim-majority kingdom. Ladakh was the north-east part of the state, a large but sparsely populated area with Buddhist majority. To the north-west were the equally sparsely populated areas of Gilgit and Baltistan. The population here was Muslim but belonged to the Shia and Ismaili sects rather than the Sunni branch. In the middle was Kashmir, including the Valley, a relatively densely populated area. It was largely made up of Sunni Muslims though it also had Sikh, Hindu and Shia minorities. Finally, to the south was Jammu, home to the Dogra Rajputs who had conquered this kingdom. The Hindu population here had increased because of the recent refugees who had come from West Punjab. Unlike Junagarh and Hyderabad, this state shared borders with both India and Pakistan. This meant that it could choose between the two. Yet, Maharaja Hari Singh dreamt of remaining independent, as some sort of Asian Switzerland. Obviously, this only made things more complicated.

And then, something unexpected happened. On 22 October 1947, thousands of armed Pakhtun tribesmen from Pakistan's North West poured into Kashmir. Nobody knows for sure who sent them or why they came there but they did have the support of the newly formed Pakistan. They made quick progress because the remote mountain valleys were cut off from the rest of the world and even Hari Singh had no idea about what was happening. He realized the seriousness of the situation only when the tribesmen blew up the Mahura power station. The entire state went dark. The invaders were now only 75 km away from Srinagar, the capital. At this stage, they could have just driven down the short undefended and well-paved road and taken over.

But the tribesmen had other plans. They decided to loot and plunder the local population, both Hindu and Muslim. They also raped the European nuns of a Franciscan mission in Baramullah, barely 50 km from the capital. They struck terror in the region for forty-eight hours and their progress to the capital was thus delayed.

The Indian government in Delhi first heard of the invasion from a very curious source. Remember, this was just two months after Independence and the Commanders-in-Chief of both the Indian and Pakistani armies were British. The Commander-in-Chief of the Pakistani army was Major General Douglas Gracey, who received secret intelligence reports of what was going on in Kashmir. The first thing he did was to pick up his phone in Rawalpindi and call his old classmate Lt General Rob Lockhart, the Commander-in-Chief of the Indian army. It didn't take long for Mountbatten and Nehru to find out what was happening.

Hari Singh panicked and signed his kingdom over to India. By the morning of 27 October, Indian troops had secured Srinagar airport and were landing men and supplies. The tribesmen had been stopped at the gates of the city. Jinnah was furious! Bit by bit, the Indians began to push back the tribesmen despite the bitterly cold winter. One of the heroes from the Indian side was Brigadier Mohammad Usman, a Muslim officer who had decided to stay in India. He was later killed in battle in July 1948.

The first Indo-Pak war in Kashmir dragged on throughout 1948. Though Srinagar was safe, western Kashmir, Gilgit and Balistan were with Pakistan. For a while, Pakistan even took over Kargil and Dras—two towns strategically important for the military. It also almost took over Ladakh. But by November 1948, Indian troops regained the two towns and secured the supply lines to Ladakh. Half a century later, Pakistan again tried to get back these towns—in what we call the Kargil War of 1999.

Some Indian commanders wanted to push ahead but they were not given permission. The matter was referred to the United Nations and ceasefire was announced. The ceasefire line is now called the Line of Control as per the Shimla Accord of 1972.

On 26 January 1950, the country became a Republic. At that time, India's borders had still not taken the shape they have today. The country had a population of 359 million—14.2 per cent of the world's population. But its share of world economy was just 4.2 per cent, compared to the 16 per cent in 1820 and nowhere close to the 30–33 per cent that it had in ancient times. The United States was now the largest economy in the world with a 27 per cent share. The Chinese economy, affected by war, was just a little larger than India's. The Chinese population was 546 million while the United States had a population of 152 million. It turned out that even dirt-poor India had a per capita income that was 40 per cent higher than that of China!

Off they Go!

So now, the British had finally left and the kingdoms ruled by princes had been handed over to the new Indian government. But there were still parts of India that were ruled by other European countries. The French had five such places. The largest was Pondicherry, south of Chennai. It was close to the old Mahabalipuram port of the Pallava kings. The others were Chandannagar (just north of Calcutta), Yanam (on the Andhra coast), Mahe (on the Kerala coast) and Karaikal (on the Tamil coast).

The French didn't want to give up these places but they knew that change was inevitable. In June 1949, Chandannagar merged with India and a year later, it became part of the state of West Bengal. The French clung on to their colonies in southern India for a few more years but finally, in 1954, they handed over all of them to India.

Pondicherry, renamed Puducherry, is today a Union Territory. That is, it is directly ruled by the central government. Most people don't know that that Yanam, Mahe and Karaikal also come under Pondicherry. The French have left but their influence still lives on. The main town of Puducherry still has many buildings from the time of their rule. The roads, planned by the French, still follow the street-grid style they had introduced. Many locals are even French citizens, descendants of people who chose to stay back at the time of the handover.

You may have heard of Aurobindo Ghosh, a freedom fighter from Bengal. He fled from the British to Pondicherry in 1910. From someone who was in the thick of politics, Aurobindo moved to spirituality and attracted a lot of followers. Though the movement has branches all over India and abroad, Pondicherry is home to many institutions as well as a commune inspired by Aurobindo.

After the French, it was now the turn of the Portuguese. The Portuguese controlled many small places along the western coast. Goa was the largest of these but there were also Diu, Daman, Dadra and Nagar Haveli. In the sixteenth century, the Portuguese had used these places to control the Indian Ocean. They were not so powerful in the twentieth century but they had managed to stick around through Vijayanagar, the Mughals and the British. They saw no reason to leave just because India had become a Republic! The Portuguese dictator Antonio Salazar declared that Goa represented the 'light of the West in the Orient'.

But the Portuguese were blind to what was happening around them. In the summer of 1954, a small group of local activists simply took over the government in Dadra and Nagar Haveli. It was not immediately absorbed into India and for a while, the area became an independent country! The Portuguese were angry. They strengthened their defences in their other territories with the help of African troops from Portuguese East Africa (now Mozambique). They used violence to put down protests and strikes. Thousands were arrested. Prime Minister Nehru had hoped that talks would help resolve the problem but by late 1961, he was getting impatient.

Operation Vijay began with the Indian Air Force bombing Dabolim airport at dawn on 18 December 1961. This is the same airport that you will land in if you fly to Goa today! Within hours, Indian ground troops were pouring into Goa from the north, south and east. The Indian Navy sailed in from the west. Similar operations were carried out at the same time in Daman and Diu.

The Portuguese planned to fight to the end but they were simply overpowered. The lone show of defiance came from *NRP Alfonso de Albuquerque*, the only Portuguese warship in Goa. Built in the 1930s, it was a medium-sized, old-fashioned warship which had no hope at all against the large, modern Indian warships. The ship was put out of service in no time but the crew still tried to fire guns from it for a while. They finally had to stop when they ran out of ammunition and the death toll on their side ran too high. The Portuguese came to India with cannon firing from their ships and they went out the same way.

Barely thirty-six hours after the battle began, the Portuguese Governor General Vassalo e Silva saw that the game was over. He signed the document handing over the territories to India. It was Christmas season but back home, in

Portugal,
Afonso de Albuquerque
as built (1934)

Lisbon, the capital city of Portugal, people were in mourning. Even the cinemas and theatres shut down! When Vassalo e Silva returned home, people gave him the cold shoulder. He was even court-martialled and then exiled. It's hard not to feel a little sorry for the man!

It's interesting to read press reports from these times about the liberation of Goa. The West seemed to think India was out of line for trying to get back territory from the European powers. The United States and Britain, in fact, tried to push for a UN resolution against India but the USSR did not agree. Many press reports spoke sadly about Goa's Christians, ignoring the fact that activists like Tristao de Braganza Cunha who fought for liberation were Christians themselves! *TIME* magazine called Nehru a hypocrite for preaching peace abroad but using violence at home. Funny they didn't notice that Nehru had waited for fourteen years for the Portuguese to come to the table for peaceful talks!

Dragon Dancing

After the French and the Portuguese, there was the Chinese to deal with. Mao's China was very powerful, not like the French and the Portuguese who were no longer at the peak of their powers.

The Sino-Indian border can be divided into two sectors. In the east, the border is called the McMahon line. It was named after Sir Arthur Henry McMahon, who was the chief negotiator for the British. The line is along the crest of the Himalayan range eastwards from Tawang, near the Bhutan tri-border and it also defines the northern boundary of the North East Frontier Agency—what is now called Arunachal Pradesh.

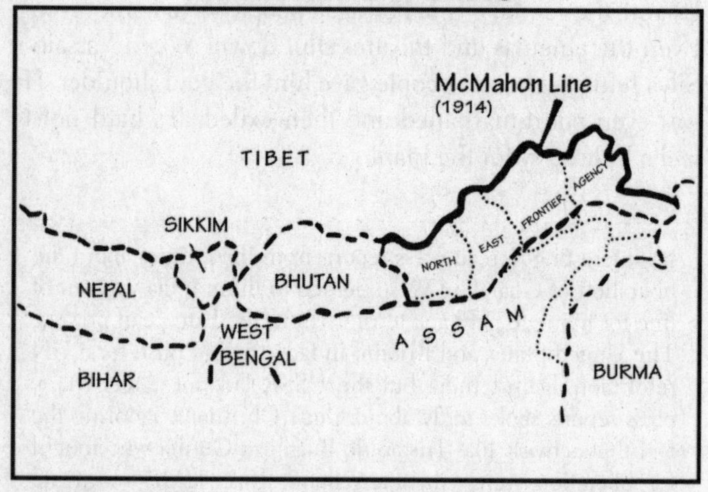

In the middle of the Himalayas, India and China were separated by three kingdoms—Nepal, Bhutan and Sikkim. The border once again continued in the western Himalayas and ran along what are now the states of Uttarakhand and Himachal Pradesh, finally running into Ladakh. Ladakh was under Indian control because it had been part of the kingdom of Jammu and Kashmir when it joined India. But who was going to control the large Aksai Chin territory that India said was part of Ladakh? Very few people lived there and it wasn't clear if it was under Indian control or the Chinese. Nineteenth-century British surveyors had drawn the borders here in two different ways.

The first one, called the Johnson Line, was drawn in 1865 between Kashmir and Turkestan. This line used the Kunlun mountains as the natural border. This meant that Aksai Chin was within Kashmir. The famous explorer Francis Younghusband visited Aksai Chin in the 1880s and reported that apart from a few groups of nomadic herdsmen and a small fort (used at times by the troops of the Maharaja of Kashmir), there was nothing much in this cold, deserted area. In 1899, the British drew a new border called Macartney-Macdonald line. This time, they used the Karakoram range as the natural boundary and left out Aksai Chin. They probably did this to create a better defence against the Russians, who they feared were expanding in this region.

The British then went on to use both the lines in their maps till 1947. No Chinese map before the 1920s showed Aksai Chin as part of China. And so, it looked like nobody really was sure who the place belonged to. The Indians probably had a slightly stronger claim.

After Independence, India's focus was on Kashmir's western border. The eastern one was unmarked and not watched. India and China were on very friendly terms in the 1950s—it was the age of 'Hindi-Chini Bhai-Bhai' which meant that the Indians and the Chinese were brothers. But to Nehru's shock, it was found in 1957 that, over the previous year, the Chinese had quietly built a highway between Tibet and Xinjiang that went right through Aksai Chin! The Indian government did not even know about it!

The quarrel between the 'brothers' began. In 1958, an official Chinese magazine published a map that showed large parts of Ladakh and the North East Frontier Agency (NEFA) as part of Chinese territory. You can now fully understand how important a role map-making plays in history! Nehru wrote angry letters to Chou En-Lai, the Chinese Premier. The Chinese responded saying that Aksai Chin had always been part of China. They did not think the McMahon Line was valid because it had been decided upon between Britain and Tibet, not the Chinese. In the middle of all this letter writing, the Dalai Lama fled to India in March 1959.

The Chinese had been claiming that Tibet had been under Chinese rule for a long time. But just as the Mughals found out in the seventeenth century, it was one thing to claim this and quite another to put it to practice. Tibet was a very cold country with a very difficult terrain, so it was not an easy job to maintain control over it. Till October 1950, Tibet was, for all practical purposes, a free country. But in 1950, it was invaded by the Chinese communists. Nehru didn't intervene though Sardar Patel did warn him about China's intentions.

By the time the Dalai Lama came to India, there was regular trouble between the Indians and the Chinese at the borders. General Thimayya, the Indian army chief, asked that the equipment be improved and that troops be sent to the Chinese border. Some units of the army were still fighting with weapons from the First World War! But Nehru and Krishna Menon, who was then the Defence Minister, did not listen.

When people began demanding that Krishna Menon resign, he promoted Brij Mohan Kaul, an officer known to be close to Nehru, to the rank of lieutenant general.

Thimayya was furious and threatened to resign. Kaul had not only bypassed twelve senior officials, he also had no field experience! On 3 October 1962, he was put in charge of defending the North East Frontier Agency.

Barely a fortnight after arriving, Kaul complained of chest pains and was sent to Delhi. And so, when the Chinese launched a full-fledged attack on the night of 19 October, the Indian troops not only lacked suffecient guns and enough soldiers, but they were also without a leader. The Chinese had attacked Ladakh, too, but there the Indian army succeeded to hold them back. In NEFA, however, the Chinese managed to take control of Tawang on 25 October. Here, they stopped for a while to build supply roads.

The Indians could have used this time to build up a more defensible position at Bomdila, where it would have been easier for them to get supplies from Assam. But Kaul insisted that the Indians should defend a position farther up at Sela Pass. When the Chinese once again started moving on 14 November, they simply went around Sela and cut off the Indian troops from behind. A large number of soldiers were killed and Bomdila fell. When this news reached Assam, there was panic. The town of Tezpur was abandoned and even the inmates of the local mental asylum were set loose. In a broadcast, Nehru said, 'My heart goes out to the people of Assam.' This sounded like he was giving up the North East to the Chinese and the Assamese people are still angry about it!

Then, just as suddenly as they had come, the Chinese decided that they were going back to where things were before the war. We still don't know why they came and why they left! The most likely reason is that winter was fast approaching and the supply lines through the Himalayas would have been difficult to maintain. In the end, it was nature that saved the Indian Republic rather than the politicians.

Today, the road from Tezpur to Bomdila is a beautiful drive through dense forests and high mountains. In the lower reaches, wild elephants often hold up traffic. From Bomdila, you can travel through the Sela Pass to the monastery at Tawang. You will see army trucks making their way up the mountain to supply the military bases that dot the region. The Chinese still mark the place as 'Southern Tibet' on their maps and made a big fuss when the Dalai Lama visited Tawang in 2009!

The war with China left thousands of Indian soldiers dead or wounded. Nehru's reputation was shattered. Krishna Menon, the Defence Minister, Lt General Kaul and army chief, General Pran Nath Thapar were removed, but it was clear to everyone that Nehru had also made huge blunders. By 1963, it was obvious to everyone that Nehru was an ageing man who had been in power for sixteen years. Once again, history seemed to be repeating itself—an ageing leader who had been on the throne for a long time. And war. The 1960s was a very uncertain period. Nehru died in 1964, Pakistan and India fought a war in 1965, Lal Bahadur Shastri (who became Prime Minister after Nehru) died in January 1966, the Congress split and the economy wasn't moving.

Out of this mess came Indira Gandhi, Nehru's daughter, the next Prime Minister. In the early seventies, she played an important role in a major shift in the political geography of the country.

The Bangladesh War 1971

The Partition of India had taken place in 1947 because of different ideas about what a nation should be like. But Pakistan faced the same problem in the 1960s. The basis of its nationhood was the idea that it was an Islamic country with an Islamic culture. But though their religion was the same, there were huge cultural differences between East and West Pakistan. In the east, there was a strong sense of being Bengali. The east Pakistanis were also angry that the political power of the country was in the hands of West Pakistan. These leaders did not care about the needs of the east. It was as if East Pakistan had simply exchanged one form of colonial rule for another.

As the Bengalis of East Pakistan began to make more and more demands, the leaders responded with violence. The West Pakistani military rulers openly stated that they thought the Bengalis were too influenced by Hindu culture. The Hindu Bengalis who continued to live in Pakistan were regarded with suspicion. Riots, often secretly supported by the government, broke out against these populations in the mid-sixties. The demands for freedom and fairness continued to grow.

Once again, it was nature that determined the chain of events. In November 1970, the major tropical cyclone 'Bhola' struck East Pakistan and killed between 3,00,000 and 5,00,000 people. It is considered to be one of the worst natural disasters on record but the military dictatorship in West Pakistan took only half-hearted relief measures to help the Bengalis. When Pakistan's military leaders finally allowed elections in late December, East Pakistan voted overwhelmingly for the Awami League, the Bengali-nationalist party. It won 167 of the 169 seats in East Pakistan.

Since East Pakistan's population was higher than that of West Pakistan's, the worry was that the Bengalis would now rule the entire country. This was bad news for the military and Zulfikar Ali Bhutto, the leader of the largest party in West Pakistan. So what did they do? They 'cancelled' the elections! East Pakistan broke out into open revolt.

The military government of Yahya Khan sent in troops. The result was mass murder, in which as many as three million people, especially intellectuals, and those belonging to minority groups, were killed. The residential halls of Dhaka

University were particularly targeted. Up to 700 students were killed in a single attack on Jagannath Hall. Many well-known professors, Hindu and Muslim, were murdered. Hundreds of thousands of women were raped. By September 1971, ten million refugees poured into eastern India. This was one of the worst mass killings or genocides in human history but people outside of the subcontinent barely know about it.

The other countries of the world behaved shamefully. The Chinese Premier Chou En-Lai actually sent a letter of support to the Pakistan government and even hinted that the Chinese would give Pakistan military support if the Indians try to interfere. The West was aware of what was happening. We now have copies of desperate cables sent by diplomat Archer Blood and his colleagues at the US consulate in Dacca (Dhaka) pleading with the US government to stop supporting the genocide. But US President Nixon was determined to keep Indira Gandhi out.

But Prime Minister Indira Gandhi began to prepare for war. Pakistan's military was in a very strong position—it had the support of the US and China, or at least their promises of support. Pakistan ordered air strikes against India on 3 December 1971. The next morning, the *Statesman* newspaper carried the headline 'It's War'. The Indian response was swift and sharp. With support from the civilians as well as the Mukti Bahini, an irregular army of Bengali rebels, the Indian army swept into East Pakistan.

It was winter and the snow-covered mountains meant China couldn't help immediately. Nixon was busy fighting in Vietnam and could do little more than make threats. On 16 December, the Pakistanis surrendered. And Bangladesh was born. But since the genocide was conveniently forgotten, no Pakistani official was ever punished for what happened and it is only very recently that some people are being punished in Bangladesh itself.

In 1975, Sikkim became a part of India. It had been ruled by the Chogyal, a ruler of Bhutiya origin who was unpopular with the Nepali people, who formed the majority. This led to constant protests and demands that he step down. The Indians were worried that the Chinese would next claim that Sikkim was part of Tibet and move in. When elections were finally held, the Sikkim National Congress, which was against the rule of the king, won all the seats but one. Later, in April 1975, the people of Sikkim voted to join India.

India still has serious border issues with China and Pakistan. Even with Bangladesh, there are issues left over from the Partition involving small regions that are still trapped in each other's territory. Still, this is how the map of India came to look like what you see today.

The Good, the Bad and the Ugly

Almost a century ago, Mahatma Gandhi said, 'India lives in its villages.' He wasn't talking about the population of the country but about the soul of India. Many people seem to think that India is mostly a rural country and that it has always been and will be so. But there have been very many cycles of urbanization over the centuries in India. And it now looks like we are in a state of rapid urbanization that will make India an urban-majority country within a generation. That is, most people will be living in cities rather than in villages pretty soon.

When India became a Republic in 1950, the percentage of people living in cities was 17 per cent. In China it was 12 per cent. The largest cities in India at that stage were Kolkata with a population of 2.6 million, followed by Mumbai at 1.5 million, Chennai at 0.8 million and Delhi at 0.7 million. The Chinese cities of Shanghai and Peking (Beijing) were much larger with 3.8 million and 1.6 million people respectively.

Though Tokyo had been seriously damaged in the Second World War, it was the largest Asian city, with a population of 6.3 million. Singapore was tiny with less than one million people and not all of them were living in the city.

The country's capital was Delhi, but Kolkata was considered to be the most important industrial, commercial and cultural centre of the country. It housed the headquarters of many of India's largest companies as well as multinationals (foreign companies that have branches in many countries). The city had lost part of its surrounding area to East Pakistan but it still had industries and factories from the British era as well as new ones set up by the Indian government, like the Chittaranjan Locomotive Works. Park Street was famous for its clubs and late-night parties—it was said to be the liveliest in Asia!

But in the late sixties, communism grew in Kolkata and with it came demanding trade unions. Through the seventies and eighties, the city was repeatedly brought to a halt by strikes against 'capitalists' (the industrialists and factory-owners), American policies, the central government and even against computers! One by one, the companies moved and with that, the art-and-culture scene in the city also suffered. There was politics everywhere, including in educational institutions. By the 1980s, the middle class began to leave in search of better education and jobs. Mumbai was now the country's new commercial capital—the city for doing business.

By the fifties, Delhi had once again become a patchwork of cities—like the Delhi Ibn Batuta had seen six centuries earlier. There was Old Delhi, including Shahjehanabad and Civil Lines. Then there was Lutyens's Delhi which was dominated by the national government. There were also many colonies which had been given to the refugees who had come to Delhi after Partition. As the capital city, Delhi also needed space for the civil servants and other government employees who had to come into the city to do their jobs. And so, new government colonies were created for them. Bapanagar, Kakanagar, Satya Marg and Moti Bagh are some of them. In the seventies, a large, new township called Rama Krishna Puram was built to the south-west.

These government areas were planned in such a way that employees would get housing according to their ranks. So if you were a civil servant, you were expected to slowly make your way up this housing ladder. This ladder was there for the military, public sector, university professors and even for the private sector. Smaller versions of it were created in the state capitals and in industrial townships.

Life in the government colonies had its pluses and minuses. Design and maintenance was poor. Painted in limewash, the walls flaked off to the touch. The doors and windows expanded and contracted with the seasons. But the houses for the more senior officials were spacious, located in a convenient area, and there were parks and other such facilities available for their use. Everyone would move up the rank at around the same time, so this meant that though people were moving homes, their neighbours more or less remained the same!

By the late eighties, the children who had grown up in such colonies began to marry across communities. Till then, the Indian middle class had been strongly linked to its origins. There was the Tamil middle class, the Bengali middle class, the Punjabi middle class and so on. It's not that they weren't proud of their Indian identity but their roots were firmly tied to where they had come from. This changed with the next generation intermarrying. Suddenly, there was a group of people whose identity came from living in such housing colonies across community lines, going through the stiff examination systems, Bollywood films, cricket etc. Their roots were not really in the place of their origin (or rather the place of their parents' origin).

In the period between the mid-fifties and the mid-eighties, there were many 'modernist' buildings that came up in the cities. Somehow, the idea of such unappealing, stark buildings caught on and architects designed buildings that were not just unfriendly to the user and difficult to maintain, but were also super ugly! And so, India, which is home to the beautiful Taj Mahal is also home to some of the world's ugliest buildings. Every major city has them—Nehru Place and Inter-State Bus Terminal in Delhi, the Indian Express Building in Mumbai and the Haryana State Secretariat in Chandigarh.

In 1950, Prime Minister Nehru invited Le Corbusier, a French architect, to design the new city of Chandigarh. Although the new city was to be built at the heart of the ancient Sapta-Sindhu and very close to the Saraswati-Ghaggar, Nehru told Corbusier to create a city that was 'unfettered' by India's ancient civilization. That is, the Prime Minister did not want the city to have any links with the past. A lot of money and material was poured into building the new city. Other existing cities were also subjected to 'master plans'. Delhi was master-planned in 1962 into strict zones according to use. But then, this never really worked. New cities like Durgapur never really took off. Even Chandigarh, an expensively built city, is not of much economic or cultural value though it's been many decades since it was built. Chandigarh was constructed according to Nehru's idea of what India should be like in the future.

The twenty-first century city that has become the face of India is a chaotic, unplanned, annoying and dynamic township: Gurgaon.

Gurgaon is Kewl

Gurgaon lies to the south of Delhi. You have already read that it was here that Dronacharya, the guru of the Pandavas and the Kauravas, is said to have lived. Though Gurgaon is near Delhi, its population was estimated to be just 3990 in 1881 and nearby towns like Rewari and Farrukhnagar had much larger populations. The British used Gurgaon as district headquarters, and the town had a small market, public offices, the homes of some Europeans and a settlement called Jacombpura. An old road connected Gurgaon to Delhi through Mehrauli. This is MG Road as we know it today.

For the first few decades after Independence, Gurgaon remained a small town in a mostly rural district. The first major change came when Sanjay Gandhi, son of then-Prime Minister Indira Gandhi, acquired a large plot of land to start an automobile company in the early 1970s. This is now the Maruti-Suzuki factory, but the project didn't take off immediately.

However, from the early eighties, a number of real estate developers (people who buy land and develop it so they can sell it for profit later), particularly DLF, began to purchase farmland along the Delhi border. The idea was to build a suburbia for Delhi's retiring civil servants. Though the Maruti car factory got on its feet by 1983, nobody really thought Gurgaon would one day become what it is now.

What changed? In 1991, India liberalized its economy.

Economic liberalization means the government reduces its regulations and restrictions so that more private businesses can participate in the economy. Liberalization in short is 'the removal of controls' in order to encourage economic development.

Around the same time, communications and information technology also improved hugely. A number of multinational companies saw their opportunity—call centres and back office operations could now be outsourced to India. Delhi was a good location for this because not only were there people who'd be able to do these jobs, it also had a well-connected international airport.

But remember Delhi was master-planned? The old planners had never really thought that such offices would come up in the future and there was simply no space for them to be set up. And so, the outsourcing companies jumped across the border to Gurgaon and began to build huge facilities for this new industry. A lot of young workers moved to Gurgaon because of this. Many of these people were the children of civil servants, public sector employees, military officers and schoolteachers.

With the young people came the malls and the restaurants. As they got married and had children, apartment buildings that were more suited to their lifestyle came up. Schools and other educational institutions began to multiply. All of this happened really fast as well. You can figure that out by looking at the lone milestone that survives on MG Road under the elevated Metro line (in front of Bristol Hotel). This is now actually the city-centre but the milestone says that Gurgaon is 6 km away!

The construction of Gurgaon was not planned. The city came up because of lack of rules and a disregard for rules when they existed. What was a sleepy town till the mid-1990s has become a throbbing city full of gleaming office towers, metro stations, malls, luxury hotels and millions of jobs. Of course, Gurgaon has serious civic problems, ranging from clogged roads to bad power supply. It's also true that if it had been better managed, it would have been a more attractive city.

But it's hard to deny the bursting energy of the place. It stands for the new India which the government is struggling to keep up with.

Slumdogs with Bite

One of the important things about new India is that the children of farmers no longer want to farm. This is true across the country. There are many reasons for this change. Literacy and growing access to television are transforming attitudes as well as aspirations. But the biggest reason is probably money. The farm economy now generates 13 per cent of the GDP and it is steadily decreasing. This has happened because of various factors. Farmers can obviously see for themselves where the money is and they do not want to invest in their land any more. Indian farming has become inefficient and its rewards are too few. The children of farmers desire other options. They want the cities.

This means that there will be more and more migration to the cities in future. Large cities will grow larger, small towns will expand and brand new cities will be built. In some ways, India is going through the same phase that developed countries have gone through at some point. Development, in the end, is about shifting people from farming for their own needs to other activities. And urbanization is how we see this process in terms of space. That is, people moving from rural farming areas to the city areas. Urban India will probably have to take in 300–350 million people over the next three decades. A huge expansion!

The explosive growth of cities like Gurgaon shows that India's rapidly expanding economy can generate enough jobs for all these people who want to move in. But how to match them to the jobs? Where will they live? What facilities will they have? This is not an easy business. Countries like China managed to do this by using very harsh social controls. How will India cope?

Most people are horrified by the kind of living conditions we see in Indian slums. The usual reaction is to think that this is a housing problem. Over the decades, we have seen many slum re-development projects that try to send slum-dwellers into specially built housing blocks, often built on the outskirts. But most of these efforts have failed. More often than not, the slum-dwellers sell, rent out or even abandon these blocks and move to a new slum!

Why does this happen? Slums don't form only because people don't have houses to stay in. A slum is an economy of its own which gives people information about jobs (inside and outside), a sense of security and a feeling of community. It's through the slum that people from rural areas learn the ways of the city and become part of it. They also provide the city its blue-collar workers—maids, drivers, factory-workers.

Slums are not really new to India. We know there were slums in Harappan Dholavira, Mughal Delhi and in colonial Bombay. Slums are not unique to India either. There were slums in New York and London in the nineteenth and early twentieth century. Indian slums are full of enterprise and energy. They are also quite safe. You can walk through the average Indian slum even at night without the fear of being harmed. How does this happen? It happens because the people living there, the migrants, don't look at the slum as a symbol of all that they can't have. Instead, they look at it as a foothold into the city. Life in a slum is definitely hard but in a fast growing economy, there is enough work for people to do so they can improve the state of their lives if they work hard, show enterprise and obey the law. This is not to say slums are awesome and that the people living there don't need any help. Obviously, they need better sanitation, health and education facilities, among others. Just that in real life, an Indian slum is not the kind of hopeless place as shown in movies like *Slumdog Millionaire*.

Rocking to Munni

A city usually expands by taking in the countryside surrounding it. In some cases, the old villages are swept away. But in most parts of India, the old villages often survive despite being surrounded by the city. Scattered across modern Indian cities, there are places where you can clearly see the borders of the old villages even decades after the farmlands surrounding them have been taken over by offices, roads, houses and shops.

Allowing the past to live on in the present is not new to Indian civilization. From the villages surrounding the city come the the cattle we often see on urban roads! They are also places you may want to visit if you are looking to buy bathroom tiles or electrical fittings. Many of these villages have become part of the city very recently but some are very old and have been within the city for many generations. In Mumbai, the villages of Bandra and Walkeshwar are located at the heart of the city but they still bear the remains of their origins.

Let's look at the experience of urban villages in and around Delhi. Roughly speaking, we can say that these villages go through the following cycle. In the first stage, the farmers sell their farmland to the government or to a developer. But they usually leave the village settlement alone. The former farmers then notice that there isn't enough living space for the large groups of workers, contractors and suppliers who have come to work on the construction site in their farmland. And so, they use the money they got by selling their farmland to build a bunch of buildings within the village settlement. These buildings are often unsafe and have poor ventilation. And they become home to the workers. Thus, the village itself turns into a slum and the former farmers become slumlords.

What happens next? The construction work comes to an end in the area and the workers move away to other places where there is a better chance of earning a livelihood. New people move into the village because there are now jobs in the newly built buildings—security guards, maids, drivers and others. The shops selling construction material and hardware change into shops that sell mobile phones, street food, car parts and so on. Facilities such as common toilets are set up. As the new batch of workers settles in, they bring along their families from their ancestral villages. English medium schools come up—the language is held by the poor as the single most important tool for going up the social ladder.

Another ten to fifteen years and the village goes through its third transformation. By this time, the surrounding area is well settled and the open fields are a thing of the past. We now see students, salesmen and small businessmen move into the village. Some of them may be the newly educated children of those who came to the area as workers, the children of migrants, but they are now of a higher social class.

The old villagers continue to be the dominant owners of the land but they also spend money to improve their properties so they can get better rent. These buildings are now in a prime location, after all. In many cases, the owners also have political connections by this stage and they manage to get drainage and sanitation facilities. The shops improve, the old street-food shops become cheap restaurants.

In the final stage, the old village becomes a place that suits the tastes of the urban middle class. This can happen in a number of ways. Since the early nineties, Hauz Khas village has become a place full of boutiques, shops, art galleries and trendy restaurants. Mahipalpur, near the international airport, has seen an explosion of cheap hotels in the last decade. Anyone driving to or from the airport would have seen the screaming neon signs that are quite like those of Hong Kong's Wan Chai district. Similarly, Shahpur Jat has become home to many small offices and designer workshops.

The old farmers now become part of the real estate business. The old farmlands now have new problems—parking space! Out of this messy process, a new India emerges. It's dominated by the new middle class who are very different from the middle class of the past. They come from slums and small towns. They are usually the first in their family who can speak some English. Their parents were probably the first in the family to become literate. Their grandparents were probably illiterate, small-time farmers. The new middle class works in call centres or as shop assistants in malls. Sports heroes in India used to come from wealthy backgrounds but now, they come from more modest social backgrounds.

As the new middle class goes up the social ladder, its tastes and attitudes change what is considered mainstream in society. You can see it happening everywhere—from Bollywood music to television news. Find yourself humming to *Munni*? Welcome to the new mainstream! The uppity people of the old middle class may not like this but this is generally a good thing.

OUTSOURCED

As we saw in the earlier chapter, Indians had once again begun to travel and settle abroad during the time of the British. What happened to these people after British rule in India ended? In some places like Singapore and Mauritius, the Indian community did well. But in many other places, they ran into problems. In 1962, the Indian community in Burma was expelled by the dictator Ne Win. Property owned by Indians was taken away from them. The same thing happened to the Gujaratis who had moved to Uganda. Some of these groups came back to India but some left for other countries. The Ugandan Gujaratis, for example, moved to Britain and became successful business people there.

After Independence, the nature of how people moved from India to other countries changed. There was one wave in the fifties and the sixties, with Punjabis moving to the United Kingdom as industrial workers. Another was of Anglo-Indians who moved to Australia and Canada. By the 1970s, the oil-rich Arab states in the Persian Gulf became a popular destination because they needed labourers in large numbers for their construction sites. Most labourers who moved to Saudi Arabia were from Kerala and many of these people were descendants of Arabs who had come to India in the Middle Ages to trade. By the 1990s, large Indian communities settled down in places like Dubai.

Historically, Indian workers who had moved abroad were mostly people who were working in low-paying jobs. Jobs which are called blue-collar jobs. From the late sixties, however, Indians were going abroad to study and to work in high-paying (or white-collar) jobs. These were the middle-class Indians. By the late eighties, it became common for Indian students to take the SAT and GMAT exams and apply to foreign universities. The United States was the country of choice for most people but many also went to Britain, Canada and other countries. By the late nineties, there was another group—Indian professionals who were hired abroad for jobs in medicine, law, finance and information technology.

Over the years, many of these groups have mixed and merged but traces of each stream can still be seen in the twenty-first century. It's interesting that many of these people, though they may never have been to the subcontinent, feel very 'Indian'. They may live in Sweden or Canada but they will probably relish snacks from India with great fondness. Many of them also feel very proud of India's economic growth.

What does it mean to be Indian in the twenty-first century? There are about twenty-five to thirty million Indians living outside India. Through hard work, education and entrepreneurship, they have become very successful in fields ranging from business and politics to literature. With success, they have become more confident about their identity. They are also able to have business, personal and cultural connections with India, thanks to globalization and communications technology.

Globalization means the coming together of ideas, cultures, worldviews and products from across the world. Eating a foreign chocolate may have been a rare treat in the time of your parents but now, you can get a bar of Toblerone quite easily!

Richer and better off socially, these Indians living abroad share passions ranging from Bollywood to cricket with their cousins back in India.

None of this is just one-way. Indians within India tend to be very proud of the personal achievements of people of Indian origin even if they have no direct link to the subcontinent. An Indian-origin governor of an American state, a Nobel Prize winner or a CEO of a multinational company can make headlines in Indian newspapers. Simply said, Indians in India and Indians living abroad have a sense of shared identity. In response, the Indian Republic has tried to create different forms of citizenship—like Overseas Citizen of India and Person of Indian Origin.

The Indian has come a long way: from the docks of Lothal to the boardrooms of London, New York and Singapore.

Put Your Hands Up, Bollywood Style

The journey from Gondwana to Gurgaon has been a long one. You may have got a sense of the twists and turns, the abrupt shifts, as well as the surprising continuities in India's history from this book. It's amazing how pieces from this long history are often piled up next to each other. For example, the brand new city of Gurgaon is being constructed right next to the Aravalli ridge, the oldest geological feature on this planet.

If you look north from one of Gurgaon's tall office blocks, you can see the Qutub Minar, built by a Turkish slave-general to mark the conquest of Delhi. Just below this tower, Indians with international tastes enjoy Thai and Italian food at the expensive restaurants of Mehrauli, an urban village that is steadily changing. Metro trains slither nearby on their elevated tracks.

A short drive south of the imperial inscriptions of Junagarh, the Asiatic lion is slowly making a comeback. A survey in 2010 reported that there are now 411 lions in Gir. The sanctuary has now become too small for these animals and some of them are wandering into the countryside surrounding it. Some have even been seen on the beaches of Kodinar!

Just across from this beach is the island of Diu, which was controlled by the Portuguese for four centuries. At some point during their rule, they had presented a group of African slaves to the Nawab of Junagarh. The direct descendants of these slaves now live in hamlets just outside Gir National Park. And so, there's one more genetic mix in India—African Indians! This community, called the Sidi community, is Muslim, but they retain customs, music and dances from Africa.

The last hundred years have not been kind to India's tigers. It is estimated that barely 1707 of them remain in the wild—down from over 3600 in the 1990s. Poaching is a big problem, but worse is the destruction of their habitat. But the symbolic value of the lion and the tiger is still alive in the minds of people. Every year, the drums of Kolkata beat for Goddess Durga, who rides a lion when she battles with evil. In Singapore, tourists take snaps of the Merlion, half-lion, half-mermaid, a mythical beast that through many twists and turns traces its origin to the ancient Indian merchants who brought their culture to this place. In 2009, the Sri Lankan army, flying a lion flag, defeated the Tamil Tigers in the long civil war.

The Indian government came up with a proposal recently to bring the cheetah back to this country from Africa. There were furious debates about whether the Asian and African cheetahs were from the same species. There are even more furious debates about mining. The ancient forests of Gondwana are now rich coal fields in Jharkhand. India's new economy needs more energy—which can be obtained from these old forests. But we really must think how we can do this without destroying the environment.

We live in a time of massive change—mass urbanization, climate change, globalization, and a changing international community. India has seen all this before but have we learnt from the past? The Ganga is still considered to be a sacred river but it's clearly dying because of human activities and thoughtless civil engineering. Maybe this is how the Harappans felt when they watched the Saraswati drying up. Maybe they desperately prayed to Indra to break the dams and let the waters flow again.

As we have seen, in spite of all these changes, Indians do have a memory of their past. It also influences our present in many ways. A lot can be known about a culture and its people from the way they remember their saddest moments. When New York observes an anniversary of the 9/11 attacks, there are serious speeches given by political leaders. But how did Mumbai remember the terrorist attacks of 26 November 2008? A day after the fourth anniversary, a flash mob of 200 young boys and girls suddenly appeared in the middle of the Chhatrapati Shivaji Terminal, a busy train station, which had been one of the places attacked on that night of horror. The flash mob then danced for five minutes to a popular Bollywood song—*Rang de Basanti* (which means 'the colour of sacrifice'). Then, when the music stopped, the mob disappeared into the crowd. In any other country, this would have been seen as an insult to the memory of those who died that night but in India, most people thought this was appropriate. The whole episode was filmed and became an instant hit on the Internet. But, why do Indians remember a horrible event by dancing?

The answer may lie in the fact that Indians view history not in a political way but as a civilization. When Americans raise their flag at the 9/11 sites, they see in it the strength of their nation. When Indians dance at the site of the 26/11 terror attack, they celebrate their civilization.

The history of India's geography and civilization reminds us of each generation's insignificance in the vastness of times. The greatest of India's kings and thinkers also felt this. So they left behind their stories and thoughts recorded in ballads, folktales, epics and inscriptions. Even if these memories are not exactly true, what matters is that they carry the essence of India's civilization. On the island of Mauritius, descendants of Indians who moved two centuries ago have a lake called Ganga Talao, named after the Ganga river, which they hold as sacred. A very long time ago, their distant ancestors would have remembered the Saraswati the same way as they shifted to Ganga. Geography is not just about the physical land but also about the meaning we give it. And so, the Saraswati flows, invisibly, at Allahabad.